But Not a Drop to Drink!

BUT NOT
A DROP
TO DRINK!

The Lifesaving Guide to Good Water

STEVE COFFEL

RAWSON ASSOCIATES • NEW YORK

Rawson Associates
Macmillan Publishing Company
866 Third Avenue, New York, N.Y. 10022
Collier Macmillan Canada, Inc.

Library of Congress Cataloging in Publication Data

Coffel, Steve.
But not a drop to drink!
Bibliography
Includes index.
1. Water-supply—United States.   2. Water—
Pollution—United States.   3. Drinking water—
United States.   4. Water—Law and legislation—
United States.   I. Title.
TD223.C645   1988   363.7'394   87-43084
ISBN 0-89256-328-1

Macmillan books are available at special discounts for bulk purchases for
sales promotions, premiums, fund-raising, or educational use.
For details, contact:
Special Sales Director
Macmillan Publishing Company
866 Third Avenue
New York, N.Y. 10022

Packaged by Rapid Transcript, a division of March Tenth, Inc.

Composition by Folio Graphics Co., Inc.

Designed by Lynn Braswell

10 9 8 7 6 5 4 3 2 1

Printed in the United States of America

Contents

But Not a Drop to Drink!

Before You Read This Book

We Americans enjoy no immunity to bad water. The dry 1980s have taught us that communities and individuals in virtually every part of the nation can expect, at one time or another, to suffer water shortages and poor water quality. Recurrent water shortages have plagued even the Pacific Northwest, the Northeast, and the Southeast—areas accustomed to a plentiful supply of water. Deteriorating water quality has accompanied the dryer-than-normal conditions as water sources dry up and pollutants become more concentrated.

The only way to ensure that *you* don't become the victim of bad water is to take definite steps to avoid it. This book explains, in understandable language, the threat posed by the toxic elements found in drinking water and what you can do to safeguard the quality of the water in *your* home.

Water is an aspect of everyday life which we normally take for granted. But the *lack* of water is something else. Withering crops; plummeting water levels in reservoirs, lakes, and streams; and restrictions on lawn watering, car washing and other normal summertime activities make it hard to forget about water problems.

During the summer of 1988, as this book went to the printer, much of the nation was in the grip of a deepening drought that by

1

midyear had already assumed crisis proportions in many areas. The first half of 1988 was the driest on record. By early July, much of the grain crop in the northern Great Plains was doomed, corn and soybeans were severely damaged in many regions, and many other crops were threatened. A drought emergency was declared in 40 percent of the nation's counties, and Congress considered a $5.5-billion drought-relief package to assist farmers stricken by the unusually dry weather. Prices for grains, corn, and soybeans were rising, and shortages were forecast if the dry weather continued.

What are the causes of the extraordinarily dry weather of 1988? Was it just a passing phenomenon, or does it reflect a long-term climatic change? What effect does the shortage of water have on drinking water quality? Are water utility and government officials doing all they should to fend off water shortages and protect water quality? These and many related questions are laid bare by drought.

The dry years affect us all in a variety of ways. Higher costs for food, manufactured goods, transportation, energy, and, of course, water are a direct result of the dry weather. These higher prices add to the cost of living and to inflation. Fish and wildlife populations suffer. And, perhaps most importantly, drinking water quality is further threatened by drought.

The drought of 1988 caused the most damage in the Midwest. Temperatures in the northern plains have averaged about ten degrees higher than normal. Rain fell sporadically across the area, but the precipitation by midyear had been too scattered to offer broad relief. Weather forecasts saw little long-term improvement in sight. The soil in some parts of the Midwest became so sun-baked that it contracted, cracking building foundations in the process. In town, empty swimming pools and brown turf were the most obvious signs of the drought. Out in the country, stunted crops and dusty fields that were never planted because there hadn't been adequate moisture stood in mute testimony to the intensity of the drought.

Other parts of the country were also short of water:

- The Southeast was much drier than usual through most of the 1980s, and long-range forecasters predicted that the 1988 drought would shift in that direction later in the summer. Falling groundwater levels, saltwater intrusion into streams and aquifers near the ocean, and record-low streams and lakes had already resulted.

● The Rocky Mountain states received 60 percent less precipitation than normal in 1988 and were plagued by below-normal snow pack and rainfall during most of the previous several years.

● The Pacific Coast was water-short, from southern California through northern Washington, with increasing frequency during the past decade. Extensive forest fires and a shortage of drinking water resulted.

The Dry '80s

By 1988, seven of the past eight years had been unusually dry:

1980: A hot, dry summer across most of the nation. Rainfall averaged 25 to 50 percent of normal in most areas. Crop and livestock damage was estimated at $20 billion, and more than 1,000 people died as a result of the heat.

1981: The dry weather continued. The Midwest, the Southeast, and the mid-Atlantic states suffered most. Rains salvaged crops in some areas.

1982: Most areas received at least normal precipitation.

1983: Another hot, dry summer. Less than 25 percent of normal rainfall was received in Illinois, Iowa, Kansas, Nebraska, Missouri, Oklahoma, and Texas. The mid-Atlantic states were also dry. Corn and soybean crops were severely damaged.

1984: The dry weather continued with more than $1 billion in damage to wheat crops in Montana and North Dakota. Crops and grasslands in central Texas continued to dry out.

1985: A year of serious brush and forest fires caused by dry conditions in California, the northern Rockies, and the northern Midwest.

1986: A drought that started during the fall of 1985 continued to grow more serious in the Southeast. The drought in Virginia, the Carolinas, Georgia, Alabama, and Tennessee was the worst in a century with less than 25 percent of normal rainfall reported in many areas. Industries were forced to close down, and many cities ran out of drinking water. More than $1 billion in crop damages resulted. Hay, soybeans, and corn crops were especially hard hit.

1987: Another hot, dry summer in the Southeast until heavy rains came in September, too late for most crops. Dry weather returned to the region in October. Millions of acres of timberland burned in Washington, Oregon, and California, and many West Coast cities were forced to adopt mandatory conservation measures.

In recent years, communities across the nation have adopted strict conservation measures to extend tight drinking water supplies. Officials in Oak Brook, Illinois, a Chicago suburb, decided to cut off services to customers who violated conservation measures such as limits on outdoor watering. The water utility encourages citizens to report illicit watering and even has a night-time patrol that spotlights backyards to check for forbidden sprinkling. Offenders are warned—once. Then their water service is disconnected. It costs $200 to reconnect it. As clean water becomes more scarce and more costly, similar stiff conservation "incentives" can be expected in a growing number of communities.

Farmers, who in many cases were looking forward to one of their first profitable years this decade, were generally hard-hit by the 1988 drought. Those fortunate enough to live in areas that received sufficient rainfall when it was needed for crop germination and growth (and those with adequate irrigation water) could hope to benefit. But in many parts of the nation, especially the upper Midwest and parts of the South and the Southeast, the 1988 drought could prove to be the last in a long series of setbacks for farmers.

Prices paid for many farm products rose as bad news about 1988's dry weather grew worse: Wheat topped $4 a bushel, up from $2.50 a year before. Corn prices reached $3.50 a bushel, more than twice the rate paid a year earlier. Crop prices fluctuated along with farm barometers—continued high atmospheric pressure means a continued escalation of the cost of agricultural staples. Falling barometric pressure and storm clouds over the Great Plains and other dry farming areas bring lower crop prices.

Although farmers were getting significantly more for their produce, only relatively modest increases in the price of food products were forecast. A box of cereal, for instance, was expected to cost only five or ten cents more as a result of the drought. This relatively small retail increase resulted from the relatively high cost of processing, packaging, marketing, and distribution of food products. The reserve of surplus grains stored around the country also has a stabilizing effect on grain prices.

Household food budgets were not expected to increase drastically as a result of the drought. Only 16 percent of an average American family's expenses are for food, and the cost of farm products is less than one-third of that. Meat prices were actually expected to decline in the short term, since many farmers were taking animals to market rather than feeding and watering them through the drought. Meat prices were down while the supply exceeded demand but were expected to soar when shortages resulting from 1988's massive sale of livestock occurred in 1989 and 1990.

The flow of the nation's three largest rivers—the Mississippi, the Columbia, and the St. Lawrence—was 45 percent less than normal during June 1988, a record low. Because these rivers carry water from about half the 48 contiguous states, their combined flow is frequently used as a gauge of total national stream flow.

The Columbia was the largest U.S. river during June even though its volume was 42 percent less than average. The St. Lawrence was second, carrying 11 percent less than normal. The Mississippi was at a record low for June, with 61 percent of its usual volume. Thousands of barges faced long delays on the Mississippi during the 1988 summer, while dredges excavated new channels through sand bars. Many shippers were forced to use more expensive rail or truck transportation as a result.

The 1988 drought brought other, less obvious effects:
- In Louisiana, the concentration of toxics in the Mississippi River's water grew steadily higher as the water level fell. The thousands of industries that dump treated wastes into the river and its tributaries rely on *dilution* to render the materials

harmless. With less water in which to dilute the wastes, the concentration of toxic elements in the remaining water rose. Decades of widespread dumping of toxic chemicals and drilling brine across much of Louisiana caused widespread groundwater contamination, and falling water tables made matters worse. Similar problems were being encountered in most drought-stricken regions.

- Electric power prices rose while power use soared as a result of increased demand for air conditioning. Hydroelectric power plants ran at reduced capacity due to reduced stream flows. Hydroplants run by the Tennessee Valley Authority, for instance, were producing only about half of normal capacity, forcing the purchase of more expensive electricity. Nuclear or coal plants made up the deficit, and even these power sources were not immune to the effects of scarce water. Vast quantities of water—almost 1,000 gallons for each kilowatt of electricity produced—are required to cool such thermal power plants.

  A 544-megawatt coal-powered generating plant in Helena, Arkansas, shut down in late June 1988 when the Mississippi dropped below the plant's water intakes, which are normally twenty feet below the river's surface. The plant uses 170,000 gallons of water per minute to recondense steam from its power turbines. Other power plants in the region were also threatened with closure because of the lack of cooling water. And low water levels made it more difficult to deliver coal, most of which is shipped by barge, to power plants.

- The population of pintail ducks was expected to reach an all-time low in 1988 and to stay at that level for several years. Declining groundwater levels caused by the drought and pumping and development across the Plains states had dried about one-third of the area's small ponds. The pintails normally mate and raise their young on the prairie potholes. Some of the misplaced birds flew to larger lakes, while others migrated farther north into Canada. The effect, in both cases, was fewer offspring. The pintails are one of about 200 species of waterfowl that live in the northern Great Plains region. All are threatened by shrinking habitat and low water levels.

• Thousands of fish were dying in Montana's Madison River because of record high water temperatures. Readings of up to 83 degrees Fahrenheit were recorded in late June and early July. Whitefish and weaker trout die at about this temperature. Prolonged exposure to 84-degree water kills most fish. To survive, many fish swim to the deeper, cooler holes. Competition for food (and vulnerability to fishermen and other predators) increases as a result. Higher water temperatures cause fish to require more oxygen at a time when the oxygen concentration in the water is decreasing. Death or stunted growth can result. The Madison is just one of hundreds of streams where fish populations were threatened by the 1988 drought. Recovery could take many years.

CAUSES OF THE 1980S DROUGHT

The jet stream is a mile-thick, 60-mile-wide river of air that normally flows across North America from west to east, ten miles above the surface. The jet stream is *moving*; air speeds of up to 200 miles per hour have been recorded. The air current normally carries storms off the Pacific Ocean and across the Rocky Mountains, wetting down most of the West in the process. Other storms are drawn up across the Plains states from as far south as the Gulf of Mexico by the jet stream's brisk passage across the continent.

A split jet stream is responsible for the drought of 1988. Instead of flowing across the Midwest as it usually does during spring and summer, the air current split into a northern branch that carried its moisture as far as Hudson's Bay in Canada, and another branch that swept across the nation's southern boundary. The two air currents rejoined over the Atlantic.

Although the condition was unusual, it is far from unique. However, meteorologists can only guess at its causes. A similar weather pattern caused the 1930s Dust Bowl. In both instances, a stationary high-pressure area centered over the Midwest helped perpetuate the pattern. It is not known whether this stalled high-pressure zone *created* the split flow of the jet stream or whether it was simply *trapped* in the island of relatively calm air between the two arms. Whatever the cause, moisture from the Pacific was, in both instances, carried to the north and south of the nation's heartland, and rain clouds from the Gulf were corralled below the southern branch.

There is no generally accepted reason for this state of affairs, but several theories exist. Some scientists say El Niño, the massive body of warm water that has appeared sporadically off the coast of South America in recent years, is responsible. The warmer water is believed to have had a profound effect on the patterns of precipitation and drought along the Pacific coasts of North and South America.

Other analysts believe that the lack of volcanic activity near the earth's equator in recent years altered the jet stream's pattern. Physicist Paul Handler, of the University of Illinois, who in the fall of 1987 predicted the drought, was the only researcher to publicly forecast the altered weather pattern, and he based his prediction on the dormancy of equatorial volcanoes.

THE GREENHOUSE EFFECT

Other scientists say that the long-predicted "greenhouse effect" is causing the current worldwide warming and drying trend. Carbon dioxide and other atmospheric gases have the same effect as the glass in a greenhouse—light passes through them readily but infrared radiation (heat waves) does not. As a result, the inside of the greenhouse can grow quite hot on a sunny day.

The "greenhouse gases" work in the same way; most solar radiation passes right through them, while infrared energy coming out from the earth is absorbed. As a result, scientists say, when enough of the gases have become concentrated in the atmosphere, we can expect the air from the ground to six to twelve miles out to slowly get warmer, becoming about six-tenths of a degree Fahrenheit warmer each decade.

The greenhouse phenomenon plays a critical role in keeping the world temperate. Carbon dioxide from natural sources has performed this vital role for millennia, keeping the average global temperature about 50 degrees warmer than it would be otherwise; the carbon dioxide does so by intercepting heat waves before they pass out of the atmosphere into space. During the era of the dinosaurs, high atmospheric concentrations of carbon dioxide kept the earth warmer than usual. The warm weather and abundant carbon dioxide fueled unparalleled plant growth. Much of the nation's reserve of coal was formed by the decay of plants that grew under this hot tropical sun.

While sun-lovers might think that such a warming trend sounds great, in reality, the results would be nothing short of devastating. A three- to eight-degree Fahrenheit increase in the average world temperature by the mid-21st century is predicted, if carbon dioxide and the other greenhouse gases continue to build up at the current rate. This temperature increase would cause:

- Hotter, drier summers and expanding deserts;
- The shrinking of lakes and rivers as the result of increased evaporation;
- A northward shift of climatic zones—after twenty years of warming at current rates, Chicago's summers would be as warm as those now experienced in New Orleans.
- The partial melting of polar ice caps, causing a one- to four-foot increase in sea level. Thousands of square miles of coastal land would be flooded. The homes of an estimated 25 to 40 million people worldwide would be threatened with inundation. New ports would have to be built. Beaches and estuaries would be destroyed. Higher water temperatures also contribute to the swelling of the world's oceans, since water expands as it gets warmer. The warming of sea water has already raised the world's oceans an average of four inches since the turn of the century.
- Longer growing seasons in some areas that are now too cool for many crops, and more water in some currently dry regions would be about the only benefits of the global warming trend.

A June 1988 report by the National Aeronautics and Space Administration brought the greenhouse effect from the realm of science fiction into the headlines. The NASA report, based on monthly temperature data from 2,000 weather stations located around the world, concluded that the build-up of carbon dioxide and other gases is *already* responsible for an alarming rate of global warming. The report's authors say there is a 99 percent certainty that their conclusions are correct.

The report stated that on a global average the warmest four years since the 1880s have occurred during this decade, and that 1988 promised to set new records. During 1987, the world-wide average temperature increased by one-third of a degree Fahrenheit, a new record. By comparison, it is estimated that the world's average temperature increased by only one-half of a degree during the century from 1780 to 1880.

It is too soon to say with certainty that specific droughts, such as the one in the Midwest experienced in 1988, are a direct result of atmospheric pollution, but the NASA study concluded that the chances of such weather are increased by the greenhouse effect.

Reversing the build-up of carbon dioxide would be difficult. The combustion of fossil fuels and the resulting production of carbon dioxide is responsible for about half the greenhouse warming. Carbon dioxide is produced by electric power plants, automobiles, trains, airplanes, ships, industrial centers, and the countless other places where coal, natural gas, diesel oil, gasoline, and other familiar fuels are used. Coming up with alternatives will not be easy.

World-wide deforestation is speeding the build-up of carbon dioxide. Forests, especially the rain forests that have been disappearing so rapidly during the recent decades, consume a tremendous amount of carbon dioxide. Healthy forests help control the gas's concentration in the atmosphere.

Three other gases account for the other half of the global greenhouse warming trend. *Nitrous oxide*, often called laughing gas but no laughing matter in this case, has spewed from generations of car and truck tailpipes around the nation. It also enters the atmosphere in substantial quantities from chemical fertilizers. *Chlorofluorocarbons*, the spray-bottle propellants and refrigerants better known for their role in the depletion of the ozone layer, and *methane gas*, which gets into the air from decaying organic matter and other sources, are also contributing to the ozone layer's depletion.

We still have some time before the more catastrophic predicted results of greenhouse warming are due. All the solutions—widespread conversion to energy sources that produce no carbon dioxide (such as solar and nuclear energy), replanting forests, and the development of transportation not based on fossil fuels—are ambitious and expensive projects that would have to be undertaken on an international basis to be effective. But that is the kind of effort that will be necessary to reverse the heat build-up. Meanwhile, we'll be doing extremely well even to level off the rate of carbon dioxide production. The world's air today contains 25 percent more of the gas than it did in 1958.

The global warming trend is expected to cause dramatic changes in inland areas such as the American plains states, southern Europe, and Siberia. Winter snows will melt earlier, causing the ground to soak up the sun's heat during a greater portion of the year. Drier soils will contribute to the general warming because evaporation will be reduced.

In such an uncertain environment, it makes more sense than ever to be prepared for problems with the reliability of your home water supply. As too many Americans have discovered, it can be fatal to assume one's water is beyond reproach just because no problems have surfaced in the past. This book will explain the basics of drinking water supply and quality and equip you with the resources you will need to find out for yourself if the water in your home is safe. And just in case it is not, water-purification techniques are outlined, and companies that supply equipment to clean up your drinking water are reviewed. Let the chapters that follow be your guide to a clean drink of water!

1 · Today's Water Crisis

*O*ur country is facing an ever-deepening, double-barreled water crisis. The purest, most easily developed supplies of usable water have already been tapped, and a growing number of users are fighting over the dwindling remains. The pollution of those remaining reserves is the other barrel of this crisis.

Alarming levels of contamination have been found in ground and surface water used by vast numbers of people. Traces of toxic materials are being found in even the purest groundwater. Lakes and streams in every part of the country are carrying a toxic burden picked up from a host of sources.

The obvious—but not necessarily the most dangerous—pollution that tainted many U.S. waterways in the past has been removed. Raw sewage and toxic discharges from industry and other sources turned many lakes and rivers into lifeless sewers during the 1950s and 1960s. At least no rivers have recently caught fire because of severe pollution, as did Cleveland's Cuyahoga River in 1969. The river became an inferno when the oil slick on its surface ignited, with 200-foot-high flames that burned bridges and curled railroad tracks into giant corkscrews.

Today's water pollution may not be as dramatic, but it is every bit as threatening. In fact, the poisons routinely discovered in our water represent an even greater threat. Many health professionals

13

say *bad water is the primary source of disease in the United States today.* And the cost of toxic tap water goes far beyond money. It is a toll levied against the health of millions of people and of entire ecosystems.

HOW'S THE WATER?

One of the first questions someone moving into an area is likely to ask is, "How's the water?" The response to this familiar query will vary widely, even in different homes within the same community, but the *real* answer is all too often, "Worse than you think."

The Environmental Protection Agency has identified more than 700 toxic compounds in samples drawn from U.S. water systems. The agency estimates that traces of between 300 and 600 organic chemicals are typically present in one source alone, the Ohio River. More than twenty cities use this river for their water supply. These same cities usually return treated waste water to the river. Some 1,800 companies today emit waste water into the Ohio River under the terms of permits dispensed by state water quality agencies.

But the Ohio is only one case in what has become a deluge of bad water. Health-threatening levels of pesticides, herbicides, industrial by-products, metals, and a host of other harmful substances have been found in both surface and groundwater in virtually every corner of the nation. Cancer, reproductive problems, maladies of the immune and nervous systems, and an overabundance of other serious physical disorders have been traced to bad water.

Even if the water that flows from your kitchen tap is clear and sweet-tasting, it may contain dangerous substances. Many toxic materials are hard to detect even with the most advanced testing equipment. Most modern water-supply systems are tested only for the handful of substances regulated under the Safe Drinking Water Act. But there are *thousands* of toxic substances in use in our industrial society, and most of them have a perverse way of finding their way into water supplies.

It is hard to know when pollution is a threat to your health, and even when contamination is obvious you may feel it is next to impossible to clear up the polluted source. This book will tell you how to be sure that you and your family don't become the victims of such circumstances. It will clarify the threat posed by bad water and tell you what you can do to be sure that the liquid of life *you* rely on isn't poisoning you and your family. You will learn how to become a survivor rather than a victim.

WATER: THE MOST BASIC OF RESOURCES

"You can live without oil, and you can even live without love. But you can't live without water."

The words of Senator Daniel Moynihan of New York underline the seriousness of the U.S. water crisis. The disruption caused by the interruption or poisoning of the water supply of a modern American city would be a disaster dwarfing the effects of even a complete and permanent cessation of oil imports to the United States. While the lack of oil would constitute a major inconvenience that would produce drastic economic and social impact, the water cutoff would be an unparalleled disaster.

Water for cities, water for irrigation, water for recreation, water for industry: All are essential. The availability of clean, abundant supplies of water is an assumption that underlies virtually every product we use and every activity we undertake.

Enormous quantities of water are needed to make the products that are part of our everyday life. Forty thousand gallons of water are required to produce the steel in an automobile, 3,000 to grow one pound of beef, 50,000 to produce the rayon for an average living room carpet, and 900 to generate a kilowatt of electricity at a coal-fired power plant. An average of 16,000 gallons of water is consumed each day for each of the nation's inhabitants. Fifty-seven *million* gallons are used each minute. Ten gallons of water were used just in the production of your copy of this book. The use of huge quantities of water is an inescapable part of American life.

The biosphere in which we live would be so much rock and dust, were it not for the life-giving touch of water. Seen from space, clouds, oceans, lakes, and streams are the predominant features of our planet. Sections of the globe where little or no water is available are barren.

The living cell is mostly water. Blood and sap are about 90 percent water. The composition of blood and sea water is very similar. Blood is a piece of the mother sea carried by all land animals that makes life outside the ocean possible.

An adult's body is about 65 percent water. A loss of 15 percent of this liquid is usually fatal. Humans require two or three quarts of water per day just to maintain the crucial liquid balance within the body. More than a gallon may be required when a person is active. Not all this liquid need be sipped, however, since our total food intake contains at least some water. Tomatoes are 90 percent water; potatoes, 80 percent; spinach, 91 percent; apples, 85 percent; and beef, 61 percent.

Even though water is an absolute necessity, most of us assign a very low value to it. This low evaluation is a primary cause of the water crisis. The potential for conserving, reusing and cleaning up water resources is enormous, but some economic incentive must be present in order to persuade consumers to take such actions. There is generally no financial penalty for polluting a water source, except in the relatively rare instance where a serious violation of water-quality laws is discovered. Clearly, we need a reordering of priorities.

WOBURN: A WARNING TO POLLUTERS AND TO WATER USERS

Bad news about bad water has surprised communities across the nation in recent years. Residents of Woburn, Massachusetts, have heard their share of such news. A series of tragic deaths and illnesses that occurred over a period of nearly two decades in this industrial suburb, twelve miles northwest of Boston, resulted in the first suit brought by victims of pollution against those they believed to be responsible for that pollution. The Woburn lawsuit was to be the first of a rush of such actions. In California alone, suits asking for more than $2.7 billion in damages from companies operating in the state were pending in 1987.

Two wells drilled in Woburn in 1964 by the city water utility were found, in May, 1979, to be polluted with the industrial solvents trichloroethylene (TCE) and perchloroethylene (PCE) Both chemicals cause cancer in animals and are suspected human carcinogens. They are also believed to cause damage to the human heart and circulatory systems, disorders of the immune and central nervous systems, and reproductive problems. As is the case with most toxic substances, fetuses, babies, and small children are at greatest risk because of their rapid growth and low resistance to toxins.

Woburn shut down the polluted wells after the contamination was detected. Much damage had already been done, however. An investigative council reported in 1985 that there had been nineteen cases of childhood leukemia in Woburn since 1969 where six would have been expected. The highest rate of leukemia in town occurred around the area served by the polluted wells.

Eight families that had experienced serious health problems during and after the years the contaminated wells were in use filed suit in 1982 against the three companies they believed to be responsible. The thirty-three complainants said the companies' negligent actions had caused six leukemia deaths, heart disease, damage to the central nervous and immune systems, and numerous other maladies among their family members.

The families had taken on a formidable task. The case was the first of its kind and promised to be complex and costly. The costs of geologic surveys, expert witnesses, and attorney's fees were all very high. Eventually more than $2 million was spent.

However, even if the Woburn families could establish that the companies had caused the pollution, they would find it still more difficult to prove that the contaminants had caused the illnesses. A $1 million settlement from one of the three defendants before the trial began helped pay for the complex litigation.

Testimony for the initial phase of the trial began in March, 1986. At issue was whether the remaining defendants, W. R. Grace and Company and Beatrice Foods Company, were responsible for the pollution of the wells.

At the end of July, the jury decided that W. R. Grace had caused contamination of the aquifer (a permeable underground layer of sand, gravel or fractured rock that holds water) by allowing employees over the years to dump solvents on company grounds and down storm sewers. The chemicals were used to clean machine parts at the company's factory, which produces food-processing machinery. The jury found Beatrice Foods to be free of responsibility for pollution resulting from illegal dumping by trespassers on its nearby property.

The second phase of the trial, in which the plaintiffs would attempt to prove that the solvents caused the illnesses in question, was scheduled to begin in September. But W. R. Grace officials agreed to an $8 million settlement before the trial continued, insisting that the payment was not an admission of guilt.

In March, 1987, Grace was indicted for making falsified statements to the Environmental Protection Agency about the quantity of toxic materials used at its Woburn plant. The company was also charged with concealing the dumping from the EPA.

The progress of the Woburn trial was monitored closely by water professionals, environmentalists, and manufacturers, all looking for a legal precedent regarding responsibility for the effects of drinking water contamination. The judge in the case said during the trial that if the companies were found to be responsible for the illnesses, the damages collected by the plaintiffs could be "astronomical." Even though the case was settled out of court, it is one more reason why polluters (and insurers) are beginning to take a closer look at the consequences and costs of contamination. If you are a home owner, the case should also serve as a warning to you of the importance of being sure about the purity of your water supply.

WATER PROBLEMS ARE NATIONWIDE

The harsh realities faced by Woburn families are unfortunately far from unique. *A river of contaminants enters America's waters every day.* The flow of this toxic stream is most vigorous in areas with industrial development and high-density populations. Moving to rural areas, however, is no solution to water quality problems. Agricultural chemicals, petroleum products, mining wastes, septic tanks, and rural industries are just a few of the pollution sources affecting the water outside urban areas. Researchers at Cornell University have found that more than 60 percent of rural wells contain unsafe levels of one or more poisons.

Toxic waste dumps are the most visible source of drinking water contamination. As many as 2,000 of a total of 50,000 toxic waste sites in the United States threaten the health of people living nearby, according to the EPA. Polluted water is the most common route by which toxic materials get into the environment from toxic waste dumps, mine tailings, and other concentrations of poisonous wastes.

More than 20,000 toxic waste sites have been suggested for inclusion in Superfund, the national cleanup program. The Office of Technology Assessment estimates it will cost *several hundred billion dollars*, $100 billion of which will be federal funds, to clean up these sites.

The eighty million tons of toxic wastes generated every year in this country weighs out to about 700 pounds for each citizen. The EPA estimates that only about 10 percent of this mountain of toxic waste is disposed of where it's supposed to be—in a registered disposal facility. The other 90 percent would yield nearly ten tons of wastes per square mile annually, were it to be equally distributed across the nation.

LOVE CANAL WAS JUST THE BEGINNING

The names of some of the areas where toxic waste contaminants have already been discovered share the kind of fame assigned to battlegrounds. Love Canal, a neighborhood near Niagara Falls in upstate New York, is perhaps the best known.

Hooker Chemical Corporation and Olin Corporation disposed of an estimated 20,000 tons of toxic chemicals in an old canal in the Falls area. The clay-lined canal was built in 1892 by a man named Love, who intended to generate electricity by connecting the upper and lower stretches of the Niagara River. The project was never completed.

The 3,000-foot long by 60-foot wide excavation that Love left behind looked ideal to Hooker officials in 1947, when they were looking for a site for another chemical waste dump. The company already had several in the same area. The site was purchased later that year and the company began disposing of chemicals there.

When the ill-fated canal was filled in 1953, the wastes were covered with a layer of clay and then soil. Hooker deeded the land to the local board of education in exchange for a token payment of one dollar. The company warned the board in the deed that the site had been used as a waste dump and assigned to the new owners all responsibility for the effects of toxic substances stored there. The board of education eventually decided to build a school and a playground on the property. New homes were also built in the developing neighborhood, near the covered waste dump.

In 1976, after several unusually wet years, water laced with deadly chemicals began to leak into basements, dissolve tree roots, and kill vegetation in the neighborhood. In 1979 it became clear that contamination in the area near the dump was causing severe health problems for the people who lived there. Miscarriages, birth defects, blood disease, epilepsy, hyperactivity, and cancer were all found in the neighborhood at alarming rates. President Carter eventually declared Love Canal a disaster area, and more than a thousand families were evacuated. The school and playground were closed and the entire area was fenced off. The tragedy at Love Canal helped speed passage of the Superfund waste-cleanup program in 1980.

The really frightening part of the Love Canal story is that most experts agree there are countless Love Canals yet to be discovered in our country. *New waste sites are discovered literally every day, typically only after health effects from the toxins escaping from the dumps alert residents of the area to the problem*. The only difference between Love Canal and the innumerable other toxic waste sites around the country is that the extent of the damage from the chemicals leaking from the old canal is known. Many of the sites now on the Superfund priority list cover much larger areas and involve more potent substances than were dumped at Love Canal.

Leaking toxic waste dumps offer just the most visible and dramatic examples of water contamination. They are only the tip of this terrible iceberg. While the extent of the pollution caused by chemical dumps, mining wastes, and other concentrated sources of pollution is wide, even more contaminants get into the water from more diffuse sources. The agricultural chemicals used by farmers and foresters to control unwanted weeds and trees, salts and nutrients carried by irrigation water into streams and aquifers, runoff laced with various kinds of petrochemical fallout, gasoline and other toxic materials picked up on city streets, and substances leaking from underground storage tanks are all part of the gathering tidal wave of pollutants.

THE LUST PHENOMENON

Leaking underground storage tanks (commonly abbreviated LUST, an acronym that's a bit zestier than most) are a rapidly proliferating source of aquifer pollution. Hundreds of thousands of storage tanks that have been in the ground since the 1960s and early 1970s are reaching the end of their lives. The EPA estimates that 75,000 to 100,000 such tanks are now leaking, and that the total is likely to reach 350,000 within five years. Many of the more than 250 components of gasoline are suspected of causing cancer and other disorders. An estimated three million gallons of gas leak from underground fuel tanks in the United States annually. Just *one* gallon of gasoline per day leaking into an aquifer can taint the water supply of a city of 50,000 people.

THE LEAD EPIDEMIC

Even when the water fed into a water utility's distribution system is of good quality, pollutants are frequently picked up en route to your kitchen tap. For instance, the EPA estimates thirty-eight million Americans drink water that exceeds the 50 parts per billion (ppb) maximum for lead now recommended by the agency. *Seventy million Americans' water carries more lead than the 20 ppb recommended by the EPA*. Most of this lead comes from lead pipes in utilities' distribution networks and from lead solder used in the newer copper home plumbing systems.

While the effects of lead poisoning on nervous system development and intelligence in children are well-known, researchers are now finding that a wide range of disorders is linked to exposure to low levels of lead in drinking water. In adults, these include fatigue, increased blood pressure, and risk of hypertensive heart disease; in post-menopausal women, accelerated loss of bone mass can result. Exposure to even low levels of lead has also been found to inhibit fetal and infant growth and to cause hearing loss in children.

A recent study concluded that 77 percent of Americans have more lead in their bodies than is considered safe. A 1985 survey of U.S. water-supply systems by Water Test Corporation found 73 percent contained at least some lead pipe. The study found that 98 percent of the plumbing in the nation's homes contains lead solder.

A frightening example of lead poisoning was found in Washington, D.C., where, as the result of a year's complaints from the parents of two-year-old twins, officials were recently forced to investigate the lead levels in the water delivered to that city's residences. The parents claimed that the lead in their daughters' blood resulted from drinking city water. They said the lead contamination had stunted the twins' growth. In the face of mounting public pressure for some response to the couple's complaints, a consultant was hired to sample water drawn from homes hooked up to the city's water system.

The results of the consultant's survey moved D.C. officials to warn residents and landlords that the plumbing systems in 71,000 of the city's houses and apartment buildings needed to be replaced to improve the quality of the water they deliver. The city is providing free blood tests for children under six to determine whether the children have suffered lead poisoning.

Lead is only one of many contaminants that may get into water as it is treated and passes through the supply system. Chlorine and other chemicals can react with impurities in the water to form toxins such as chloroform, which is a carcinogen. Plastic pipes in the distribution system are vulnerable to permeation by toxic compounds, including most petroleum products. If plastic water mains pass through a contaminated area, pollutants can enter the water they carry. Corrosion by-products and contaminants that enter though breaks in supply systems can also degrade water quality.

IS THERE HOPE FOR IMPROVING OUR WATER SUPPLY?

Is there reason for hope in this decade of bad and dwindling water? Are steps being taken to improve our water? More importantly, what can *you* do to be sure your water is not slowly poisoning you?

GOOD WATER IS A MAJOR PUBLIC CONCERN

First the good news. In response to the water crisis, a growing reservoir of assistance is becoming available. Testing services, purification systems that will remove everything but the "wet" from your water, consultants who will figure out how contaminants are entering your water supply, and bottled water distributors are all increasingly available and effective—if you know how to choose wisely among them.

Another piece of good news is the bad news: All the negative reports about water have Americans upset. In a 1986 Harris Poll, 86 percent of those surveyed agreed that there should be no exceptions to standards for water pollution even if lost jobs and financial hardship on the company owning the factory in question resulted. Seventy-three percent of Californians consulted in a 1986 University of Southern California poll said they did not trust the government to protect them from the effects of toxic chemicals. This growing public concern means that water utilities and regulators, even the many reluctant ones, are being forced to take a new look at water treatment and distribution systems.

Politicians, too, are aware of the public's insistence that drinking water supplies be clean. Most of the federal legislation designed to protect the quality of drinking water was renewed during 1986 and 1987, after years of debate. A revised Safe Drinking Water Act, passed in May, 1986, will require water utilities to take a closer look at the quality of the water they supply. The law established tighter controls for contaminants found in drinking water and required monitoring for a total of eighty-three pollutants, rather than the current twenty-two.

A $10 billion Superfund renewal bill was signed into law by President Reagan that same year, despite his threatening until the last moment to veto the bill because he said it cost too much and put too many restraints on the EPA's administration of the program.

Overwhelming approval of a $20 billion renewal of the Clean Water Act, which will provide funds for local governments to build sewage treatment plants, was a first order of business in Congress in 1987. President Reagan, who pocket-vetoed virtually the same bill after Congress adjourned for Christmas in 1986, rejected bipartisan appeals to sign the legislation, but his veto was overridden.

In many states, voters have chosen not to wait for federal programs to clean up their polluted water. Initiatives calling for toxic waste cleanups passed by large margins in several states. In California, Proposition 65 passed by a two-to-one margin. The measure calls for an accelerated cleanup of toxic wastes in the state and prohibits *any* contamination of water supplies with substances known to cause cancer or reproductive toxicity. An industry wishing to dispose of wastes must prove that the discharges will be harmless before a disposal permit will be issued.

In Massachusetts, 73 percent of the voters endorsed a toxic waste initiative intended to start cleaning up sites not covered by Superfund. The measure rolled up the widest margin of support ever recorded for an initiative in that state. Two-thirds of the voters in New Jersey backed a $200 million bond issue to clean up toxic waste sites. Residents of New York passed a $1.45 billion environmental bond issue for the same purpose. Legislatures in several states have passed or are considering similar measures.

The efforts being made by some cities and polluters to clean up contaminated water supplies is also good news. Modern treatment and testing techniques are being combined with time-proven practices to remove contaminants from waste water. Earth, sand, and growing plants are being used to clean up water, as they were a century ago, with modern instruments, materials, and methods improving the efficiency of the treatment process. Old water mains are being replaced to reduce the loss of good water and the infiltration of contaminants into the supply system.

PURE WATER IS COSTLY, BUT A GOOD BARGAIN

The bad news: Getting pure water won't be cheap. In a nation where abundant supplies of water have long been taken for granted, this is bad news indeed. We have had so much water at our disposal across most of the country for so long that we often have trouble acknowledging that pure, abundant water is a thing of the past. The out-of-sight, out-of-mind philosophy of waste disposal and the assumption that we will always have plenty of water are still alive and well, even in the face of incontrovertible evidence to the contrary. Paying the real cost of providing reliably clean water, whether for a townhouse or a town, will take some getting used to, but it is a cost that must be paid. The $23 billion spent in this country to clean up water during 1983 will undoubtedly be dwarfed by 1993 statistics.

The fact is, however, that cleaning up our water is probably one of the best bargains around. Compared to even the direct costs of the alternative—using water that is unhealthy—it makes good economic sense. If spending money on testing and cleaning the water in your home can save a member of your family from a debilitating disease, is there any better way to invest the money?

If the local water utility must double its rates to ensure a level of quality that will result in a general improvement in the health of its customers, isn't the expense justified? An ounce of prevention in the form of paying for good water quality is indeed less expensive than a pound of cure.

This book outlines—in terms easily understood by the lay person—a definite series of steps *you* can take to apply that ounce of prevention. You will learn how to assess your water's quality and how to purchase, install, and maintain the appropriate purification equipment. You'll even learn how to select bottled water that's of good quality (some bottled waters contain as many poisons as tap water—because that's exactly what goes into the bottle).

Chapter 8 is a guide to guaranteeing the quality of the water in *your* home. More than 200 toxic chemicals are listed in a table that summarizes the threat to health they represent and the symptoms associated with exposure. The chapter also indicates how best to remove each from your water, where the materials are likely to be found, and what they are used for. Laboratories that can test your water and sources of water purification equipment are also reviewed in chapter 8. It is your guide to reliably clean water.

In the midst of all the doom and gloom about bad water—and there are enough horror stories to fill many a volume—there is reason for hope. A few individuals and communities are taking dramatic steps to improve the quality of their water. In chapter 8 you will learn how families across the nation discovered their home's water was unsafe and how they got rid of the toxic compounds it contained. In chapter 11 you'll learn what some of the country's more progressive water utilities are doing to upgrade the quality of the water they supply.

Although removing the impurities from water requires no technical background on the part of the homeowner, the process is far from simple. The water-filter salesman who tells his potential customers that one filter will remove *everything* objectionable from their water is not telling the truth. The thousands of water-borne contaminants that have been found in U.S. water supplies come in all shapes and sizes; a purification device that works perfectly for one contaminant will prove inadequate in dealing with others. Even finding out which undesirable elements are present in your water supply can be like looking for a needle in a haystack—without the necessary guidance. Testing the water for *all* impurities is prohibitively expensive. This book will tell you how to get a better idea of what the toxic needle-in-a-haystack you want to remove looks like, and how best to remove the pollutants. With this information in your possession, you will be spared the often-expensive trial-and-error approach to cleaner water.

2 · No Water

DISAPPEARING WATER

An alien viewing our aqua-green planet from space would never guess that the majority of the earth's five billion human residents do not have access to a reliable, clean supply of water. More than 70 percent of the world's surface is covered with water up to seven miles deep. The water bound to the earth would form a sphere with half the diameter of the moon. But still, there isn't enough water for everyone. Aquifers are rapidly being depleted, and most surface streams are already overused. Pollution further limits these dwindling supplies.

Water use in the United States increased from 180 billion gallons a day in 1950 to 450 billion gallons a day in 1980, according to the U.S. Geological Survey. This is a frightening two-and-a-half-fold increase in just thirty years. Irrigation use increased from 89 billion to 225 billion gallons a day during the same thirty-year period, and the water used by public water-supply systems jumped from 14 to 34 billion gallons. This astonishing increase in water use has put many areas' water budgets into the deficit column. With similar growth forecast in demand for water over the next few decades, existing resources just will not be able to do the job. Something has to give.

What often gives first is water quality. Despite the well-established habit of dealing with water quality and supply problems piecemeal, the two subjects are intimately related. They are flip sides of the same coin, sharing some common roots and some common solutions. The general approach is to expand water supplies to meet growing demands; expand waste-water treatment facilities to keep pace; and later to expand pollution programs to deal with the contamination that has been an inevitable part of growth. Unfortunately, poor water quality is the result.

Streams that no longer have an adequate summer flow to support fish, streams that *disappear* altogether (as does the Colorado River before completing its ancestral journey to the Gulf of California), depleted aquifers, devastated wetlands, shrinking lakes: All these are manifestations of the water supply crisis that threatens to become, in some areas, the *no*-water crisis.

In the Red: The World's Water Budget

The earth's total allotment of water has a volume of about 344 million cubic miles. Of this:

- Three hundred fifteen million cubic miles (92 percent) are sea water.
- Seven million cubic miles (2 percent) are frozen in polar ice caps.
- Nine million cubic miles (about 2.5 percent) are under the earth's surface.
- Fifty-three thousand cubic miles (less than .002 percent) are passing through the world's lakes and streams.
- Four thousand cubic miles are atmospheric moisture.
- Thirty-four hundred cubic miles are locked within the bodies of living things.

One three-thousandth of the earth's water evaporates each year, primarily from the oceans, to produce 1,500 cubic miles of rain water. Evaporation produces pure water. When a molecule absorbs sufficient energy to attain escape velocity from the body of water of which it is a part, it must travel light. Impurities are left behind. Evaporation is thus the source of fresh water's freshness and salt-water's saltiness.

Ninety-six percent of the nation's fresh water is contained in aquifers. Half the drinking water and a quarter of the water used for all purposes is groundwater. Most aquifers are too deep for practical use. More than two-thirds of the groundwater underlying the United States is estimated to be 2,000 or more feet below the surface.

Agriculture accounts for the bulk of our national water use. The average American consumes 1,500 pounds of food each year. One thousand gallons of water are required to grow and process *each pound* of that food. This means that in this country in a single year an average total of one-and-a-half million gallons of water is invested in the food eaten by *just one person*. This 200,000-cubic-feet-plus of water would be enough to cover a football field four feet deep.

Seventy-five percent of the water used for irrigation is consumed through evaporation from the surface of the ground and from the leaves of plants. An average of only 11 percent of the water withdrawn for industry is actually used up. The rest flows back into streams and lakes from discharge pipes or into aquifers via seepage and injection wells. Domestic and commercial uses of water consume an average of 22 percent of the total that is withdrawn. Only 0.6 percent of the water used for cooling electric generating plants is consumed. Twenty-five percent of the total cash value of the nation's crops is produced on irrigated farmland, which comprises 10 percent of the total farm acreage.

THE ENORMOUS COSTS OF WATER DISTRIBUTION

The uneven distribution of water is another fact that must be reckoned with. Although basically abundant, U.S. water is poorly dispersed, both in time and space. A season of torrential downpours followed by one of drought is an often-repeated pattern. Windward slopes that receive heavy precipitation and support lush vegetation are commonly adjacent to deserts located in the mountains' rain shadow. These deserts may go years between rains and average only a few inches per year, which may come all at once, during a cloudburst.

The average of two and a half feet of precipitation that falls on the United States (not including Hawaii and Alaska) each day results in 18,000 gallons for each inhabitant. However, 40 percent falls east of the Mississippi, on 24 percent of the nation's land. The average in the Western states is just fifteen inches, and most of that falls along the coast.

Water is often not only scarce *where* it is needed; it is often absent *when* it is most wanted. Successive floods and shortages of water make providing a year-round supply an uphill battle. Forty to seventy percent of the water that flows through American rivers comes during the two to three months of flood flows in the spring. Seventy percent of the surface water in the American West is melted snow.

Thousands of reservoirs have been built to even out the seasonal flux in runoff (the water that flows into lakes and rivers from rainfall) at a federal government investment of billions of dollars. Agricultural development and the growth of urban centers in dry regions were made feasible, in many cases, by such federally subsidized water projects.

Massive water loss through increased evaporation has been one negative result. Much of the water that once rushed to the ocean each spring is now held in reservoirs through the hot, dry summer and fall. Since the surface area of the man-made lakes is much greater than that of the rivers they replaced, the torrid summer sun vaporizes more of the water.

In Colorado, for example, *more water is lost to evaporation than is used for all purposes inside the state*. The water held behind Hoover Dam in Lake Mead made phenomenal growth possible in southern California, but it is estimated that evaporation causes the loss of one cubic mile of water from the lake each year.

People are not distributed evenly either. Urban areas that are home to millions are separated by regions that are comparatively lightly populated. Local water resources cannot begin to satisfy the water demands of a megalopolis. And many population clusters are located in dry regions. The Los Angeles watershed by itself, for example, receives enough precipitation to support only 100,000 people. But *twelve million* people live there. It shouldn't come as a total shock, then, that half the total electricity used in California goes to pump water over mountain passes to Los Angeles and other coastal California population centers.

The devastating 1986 drought in the usually wet southeastern United States shows that water shortages can have severe impact anywhere. Billions of dollars in crop and livestock losses (and tens of thousands of farmers pushed over the financial brink) were only part of the cost of that drought. Pumps were left high and dry by falling groundwater levels; industries shut down when their water supplies dried up; barges quit running when canals became too low to navigate; fish died as a result of record-high water temperatures and record-low water levels in streams. Cities were forced to ration water during the drought and are now investing in expanded water-supply systems as insurance against a recurrence. Many water systems in the region have minimal storage reservoirs because of the usually dependable rainfall in the area.

SINKING LAND AND GROWING DESERTS

The use of groundwater has increased dramatically over the last few decades. With lakes and streams making up less than a twentieth of the nation's total stock of water, attention naturally turned to aquifers for irrigation and municipal water supplies when surface sources became harder to find. The flourishing agricultural regions from the Midwest and the plains of Texas to the otherwise arid valleys in the southern half of California owe much of their verdure to the humming pumps that bring water to their surface. Many aquifers are being "mined" as a result; unfortunately, more water is being removed than is seeping back in from precipitation and irrigation.

The Ogallala Aquifer, a giant reservoir of groundwater that stretches from central Texas to northern Nebraska, has been especially hard hit. The worst problems are on the Texas High Plains, where a billion-dollar agricultural industry has been built around irrigation with groundwater, in an area without rainfall or surface water to support it. About a quarter of the nation's cotton and grain sorghum and 15 percent of its feed cattle are raised there.

Water levels in many parts of the plains have already dropped so far that pumping is prohibitively expensive. Too much electricity is consumed to justify pumping groundwater from much more than a few hundred feet below the surface. Irrigated agriculture in the region will dry up if no alternative to depleting the aquifer is found, and there are no available supplies of surface water anywhere close that could provide the volume of water required to rescue the region.

Under the right geological conditions, falling water levels in overused aquifers can cause the ground above them to sink, a phenomenon called *subsidence*. In Arizona, the level of the desert between Phoenix and Tucson has done just that, dropping *fifteen feet* as the result of pumping for city water supplies and irrigation. Cracks up to nine miles long have opened up on the ground's surface as a result of the subsidence.

If current rates of subsidence continue in Houston, the top of a forty-five-story building will have sunk from sight under the Gulf's surface by the year 2180. The city's elevation has already decreased by ten feet owing to the pumping of water from the aquifer that underlies it. Mexico City has fallen thirty-five feet as the result of seventy years' pumping, even though only 15 percent of the city's residents have running water in their homes. The surface of California's San Joaquin Valley has subsided up to thirty feet in a relatively short time span, mostly as the result of irrigation.

Desertification is another fate often suffered by arid lands, partly as the result of falling water tables. Up to 35 percent of the world's land surface is threatened, an area inhabited by 850 million people. In an area undergoing desertification, the land gradually becomes less capable of supporting life as the number of native plants, animals, and soil microorganisms decreases. Many desert plants can no longer survive when water tables (a water table is the surface of the aquifer) fall beyond the reach of the long roots these plants have developed. A drought often helps accelerate the process, but there is no recovery of productivity when normal precipitation returns because of the general decline in biological activity in the area's soils. Many parts of the Southwest are currently undergoing desertification.

THE CONSERVATION QUESTIONS ARE HARDER TO ANSWER

Inadequate supplies of water and often-conflicting plans for their use can lead to hard questions. Should this water be used to vitalize a large city, or would it be more productive if used to grow a crop? Should it be left in the stream for the convenience of fish and fishermen? Or would it play a more vital role if used to slurry coal down a pipeline? Should water-rich areas (such as the Pacific Northwest) sell water to water-short areas (such as the Southwest) via long-distance diversions? Which should be sacrificed, water quality or jobs? Such questions have too often been answered with convenience and short-term gain as the determining factors.

More efficient use of existing supplies can provide a partial answer. Repairing leaky pipes and canals, installing meters, and raising rates to encourage conservation are first steps frequently taken by utilities where further expansion is too expensive or simply impossible.

A utility can also save water by promoting conservation among its customers. Outdoor watering can be reduced; leaks in home plumbing systems can be repaired; consumers can be urged to install low-flow faucets and showerheads; fixtures such as toilets that use a lot of water can be replaced by more efficient units; and more efficient plumbing systems can be encouraged in the building of new homes. Such measures can produce significant savings, all at a cost much lower than developing new sources.

A 30 percent reduction in the water used just for flushing toilets would cut typical residential water use by about 10 percent. The average flush in the Scandinavian countries uses one-third of the five gallons consumed in this country. And just fixing the leaks in the typical American water guzzler can drastically improve its efficiency.

Manufacturing businesses also have an increasing number of reasons to reduce the amount of water they use, and the volume and toxicity of waste water they generate. The growing cost of being responsible for pollution is probably the most important. Expensive court cases and settlements and the increasingly stiff fines imposed by government water-quality agencies are forcing some polluters to clean up their act. Companies are cutting water use drastically while reclaiming valuable materials.

An Armco plant in Kansas City, Missouri, for example, manufactures steel bars from ferrous scrap using only a little more than 300 cubic feet of water per ton of steel produced. Ten to twenty times this use is typical. Water in the plant is reused sixteen times before it is discharged.

Conservation not only reduces the need for additional water sources, but also improves quality. Removing too much water from a stream or lake causes the temperature of the remaining water to rise, harming native fish and plants and encouraging the growth of inferior species. Greater vulnerability to pollution also results, because the meager flow is unable to adequately dilute contaminants. A body of water receiving excess nutrients from sewage plants, farms, industrial discharges, or other sources while being overdrawn can lose most of its dissolved oxygen, with disastrous results for the aquatic environment and water quality. According to the EPA, 68 percent of the nation's streams and lakes are suffering the effects of overuse.

Severe drought can spark surprisingly effective conservation efforts. A 65 percent reduction in residential water use was achieved in the suburbs of San Francisco during the record-breaking drought of 1977. Outdoor watering was restricted, so many residents used "gray water" (water already used for other purposes) from sinks and tubs to maintain their shrubs and plants. Saving water became a subject of public enthusiasm and debate. Water use returned to normal, however, when the two-year drought ended amid flooding and mud slides.

Mandatory and voluntary restrictions on water use in New York City during the 1985 water shortage cut use by one-sixth; use dropped from an average of 1.5 billion gallons a day to 1.25 billion. (Mayor Koch's goal was to cut daily use to 1.1 billion gallons.) The reduction came during a season when water use would normally soar to 1.8 billion gallons daily, partly because during hot weather an estimated 1,000 fire hydrants are opened by people wanting to cool off or wash their car under the 1,000-gallon-per-minute geyser. The city installed locks on 30,000 hydrants located in neighborhoods where they were traditionally opened, thereby saving an estimated 400 million gallons a day. Lawn watering was also banned. Officials estimate that 15 percent will be trimmed from the city's water use after meters are installed on the 600,000 buildings that now have none, a move city officials hope to complete over the next decade. The city's 110,000 commercial and industrial water users are already metered.

Water problems are seldom easy to clarify. Figuring out how a source is being contaminated or deciding how much water can be taken from it without harm can be complex problems. Only by planning—going beyond simply reacting to trouble as it surfaces—can a reliable supply of healthful water be assured. Relying on unlimited expansion to provide ever more water must eventually lead to disaster.

DOWN ON THE FARM: SELLING THE WATER, LOSING THE LAND

The high cost of new water sources for cities is also encouraging conservation among irrigators. Some irrigation districts are saving water by reducing seepage from ditches by lining them with waterproof membranes or concrete, and encouraging conservation among irrigators to free up water to sell. On the negative side, however, many farmers, facing tough times, are selling part or all of their water rights—or even their entire farms—to city water utilities.

Aurora, Colorado, for instance, spent $50 million during 1986 to buy water rights from 300 sources in the Arkansas River Valley, primarily farmers, and thereby increased by 30 percent the water available to the city. The rights to 60,000 of the 320,000 acres of irrigated land in the southeastern Colorado valley where the Aurora water department was shopping have already been sold. Most of the land's value is lost when irrigation rights are sold, and many local residents believe that the area's rural economy has been ruined in the process—but sellers usually are financially strapped and have few alternatives.

An acre-foot represents the amount of water it takes to cover one acre one foot deep—326,000 gallons. A typical farmer in the area may have rights to a couple of hundred acre-feet of water, for which he or she may receive from $1,500 to $2,500 an acre-foot. Nationally, prices paid for water rights range from as little as $100 to as much as $10,000 per acre-foot. The price of water is likely to be discussed more often in the near future than the price of oil or gas.

THE REUSE AND RECYCLING OF WATER

Innovative solutions often solve more than one problem at once: Taking the "waste" out of waste water through reuse and recycling and stretching existing supplies through improved efficiency can save a utility the expense of developing new resources while taking some of the burden off streams, lakes, and aquifers. Recycling the waste water in a factory not only can reduce demand for fresh water, but also can eliminate pollution. The factory owner can thus avoid the growing threat of a lawsuit from downstream users of the water into which the wastes are discharged. Valuable chemicals and by-products can be reclaimed in the process; these are often the same materials that would cause damage, were they to enter the environment. Comprehensive management of water resources can solve problems that appear unsolvable through traditional approaches.

An increasing number of cities are finding that it is easier to augment their water supply by protecting existing resources and making more efficient use of the water already available. In Tucson, Arizona, per capita water use declined by one-quarter over the last fifteen years as a result of conservation efforts by the city. Tucson's sole source of water until the Central Arizona Project's canal reaches there in 1991 is groundwater. The surface of the water in the aquifer underlying the city is dropping two to ten feet a year.

Tucson saved $45 million by using conservation and recycling to avoid the planned expansion of its network of wells and pipelines. A boost in rates in the late 1970s started a reduction in personal water use. Restrictions on lawn watering and the promotion of desert landscaping as an alternative to green lawns are also responsible for a large part of the savings.

The city further stretches supplies by using seven million gallons a day of highly treated effluent (waste water) from a tertiary sewage plant built in 1983 to water parks, golf courses, school grounds, cemeteries, and other turf. The watering system's capacity will increase to thirty million gallons a day by 1995. A pipeline that will circle the city to deliver the irrigation water will also be completed by that time. The city sells the reclaimed water to private users at a bargain rate to encourage its use. City officials say that after the system reaches full output, revenues from water sales will pay operating and maintenance costs.

In El Paso, Texas, since 1985, highly treated waste water has been injected into the aquifer from which the city's drinking water is drawn. It is the first project of its kind in the United States. The waste water is cleaned up through additional filtering and chemical treatment after conventional processing is completed.

In Orange County, California, fifteen million gallons per day of treated waste water have been injected into groundwater adjacent to the coast since 1976. The water is being pumped into the aquifer to prevent saltwater intrusion resulting from the depletion of the county's primary fresh water source. Although not used directly for drinking, the water meets federal standards for drinking water, according to utility officials.

In Denver, one million gallons of water recycled from the city's sewage treatment plant are mixed into the water-supply system each day. The $30 million treatment plant operated by the city will be the forerunner of a 100-million-gallon-per-day recycling plant. That new recycling plant will supply 15 percent of the city's water needs by the year 2000, if no problems are encountered in the pilot project. Since new diversions of water to the city from the upper Colorado River basin across the Continental Divide cost more than $3,000 per acre-foot, recycling is a move that is saving money for the utility's customers. It currently costs $2.50 to treat 1,000 gallons of water at the demonstration recycling plant, a cost of a little more than $800 an acre foot. Per-gallon treatment costs are expected to be considerably less at the larger plant.

Treated outflow from the city's sewage plant is piped to the plant, where it is run through sand and coal filters, aerated with ozone gas to kill viruses and bacteria, and forced through a reverse-osmosis filter to remove single-molecule-sized contaminants. Finally, it is cascaded through an air-stripping tower to remove dissolved gases, and treated with chlorine dioxide to remove harmful by-products of chlorination. The majority of Denver residents were initially apprehensive about drinking water that had already gone down somebody's drain. When it became apparent, however, that the quality of the treated waste water is at least as good as that of the average water in the city's supply system, the concern of most residents abated.

Denver is far from being the first U.S. city to recycle waste water. San Francisco started irrigating Golden Gate Park with treated sewage effluent in 1932, at a time when such practices were widespread. Many cities have used recycled water to irrigate golf courses and other municipal greeneries.

In Chanute, Kansas, in 1956, a severe drought forced the city's water utility to mix recycled sewage effluent with fresh supplies to extend a very limited supply of water. Conventional treatment was used to clean the water. No negative health effects were reported during the five months the recycled water was used.

As mentioned in chapter 1, it takes enormous amounts of water to produce the products and carry on the activities that are a part of our everyday life. Expanding water-supply systems is a solution to dramatic increases in demand only when ample resources are available. That time has passed. Better management of existing resources becomes a necessity when aquifers and surface sources become depleted or polluted and additional water becomes too expensive to buy and transport. Far more will have to be done if the *low*-water crisis is not to become the *no*-water crisis in many parts of the country.

3 · Bad Water

*I*t seems incredible, given the depth of America's technological resources, that the water crisis has been allowed to grow to the menacing proportions we face. It's not that difficult to remove pollutants, once we know that they are there. Preventing the contamination would appear, in hindsight, to have been relatively simple. But too few were aware of the problem. Neither the sensitive testing equipment needed to detect often-minute quantities of harmful substances nor adequate information on their effects was available during the 1950s and 1960s. Public concern about severely polluted surface water during the 1970s prompted a cleanup of the obvious contamination in the nation's lakes and rivers. But even today, with dramatic evidence of the extent and health effects of pollution readily available, the flow of toxic materials into our water continues unabated.

In a sense, we are victims of our standard of living. The United States has become one of the world's leading industrial powers, but too frequently this position has been maintained through the depletion and abuse of natural resources. The technological genies that have always been called on to solve our problems are, as it turns out, the *cause* of many of those problems. The consumption and pollution of water and other resources is an integral part of modern technology.

We are also picking up the tab for innumerable environmental shortcuts. The seeds of the Love Canal disaster sprouted in the 1970s, but were sown more than twenty years earlier. Hooker Chemical and others no doubt saved a tidy sum through the expedient disposal of wastes at chemical dumps like the one in Love Canal. But the savings were short-term. The bitter fruits of such dumping will haunt for decades both the dumpers and those who later became its victims.

This kind of false economy is still considered to be good business, in many circles. Until the recycling of most industrial chemicals and the safe disposal of hazardous wastes are seen as an unavoidable part of doing business, the vigorous flow of toxic wastes into the nation's water supplies will continue.

And until the flood of toxic wastes and myriad other contaminants is dried up, bad water will continue to be a threat to our health. This chapter will describe the primary sources and extent of the contamination that is increasingly turning good water into bad, and will provide details about toxic materials *you* should be concerned about.

THE POISONS IN YOUR PIPES

In 1962, when Rachel Carson's *Silent Spring* was published, the chemical industry tried to stop its publication, implying that the book was part of a communist plot to undermine the country's vitality by causing the public to doubt the safety of essential chemicals.

The book had the effect feared by industry officials: It generated some of the first ripples of what would become a tidal wave of public concern about the impact of toxic chemicals on the environment. Today Carson's book is primarily remembered for pointing out the effects of pesticides on birds, but she was also one of the first to describe the association between the introduction of pesticides and herbicides during the 1950s and an increase in human cancer rates.

Her concern about the toxic chemicals that began their insidious infiltration into the nation's water during the 1940s and 1950s has turned out to be prophetic. Although modern water systems have eliminated the tragic epidemics that used to spread through drinking water sources, our modern world is being visited by a new kind of plague: up-to-date diseases such as cancer, reproductive toxicity, and immune and nervous system disorders. Bad water is one of the principal culprits in the spread of this modern epidemic. Today, even those with little interest in environmental issues are concerned about toxic materials in the air and water.

Incredibly, most of the chemicals attacked in Silent Spring *are still being used in this country.* For those few that have been banned, alternatives that are even more toxic have too often been substituted. Carson wrote that dieldrin, for example, is five times as potent as the banned DDT when swallowed and fifty times as potent when absorbed through the skin. Traces of dieldrin are found today in the tissue of 99.5 percent of Americans and in 96 percent of the meat, poultry, and fish consumed in the country.

The indestructibility of many of these toxins adds to the danger they pose. DDT, although banned now for more than a decade, is *still* found in practically all human tissue as well as that of most animals, fish, and fowl! Even when concentrations of a poison are low enough so that no immediate health threat is posed, dangerous levels can eventually accumulate. Once on the loose, such persistent chemicals remain a danger for decades.

Pesticides and industrial wastes are joined by pollutants from a multitude of other sources in the nation's drinking water. Diesel fuel, gasoline, motor oil, and other petroleum products have been found in water supplies virtually everywhere. Leaking underground storage tanks that have polluted aquifers with fuels and toxic chemicals can also be found universally. Solvents and degreasers containing potent carcinogens have similarly contaminated water in every part of the country. Sewage plants and municipal dumps are yet another part of our lives that routinely pollute water. Radioactive materials often are found in areas where uranium is mined or processed. Heavy metals from mining and industry also find their way into our water supplies.

THE MAJOR CAUSES OF BAD WATER

The ways in which water can pick up contaminants are as diverse as the uses to which it is put. Here are the major causes of bad water:

POLLUTED RUNOFF

The water that drains from the country's farms and forests is often laced with pesticides, herbicides, and excess nutrients and salts. This contaminated runoff does an estimated $2.2 billion damage annually. Runoff from city streets frequently contains petroleum products, lead, toxic chemicals, and even raw sewage. This polluted runoff flows into lakes and streams and—too commonly— ends up in our drinking water. Some of the sources of polluted runoff deserve special mention, as follows:

Pesticides
Almost 50,000 products containing more than 600 potentially dangerous chemicals are used to kill weeds, insects, rodents, and other pests on our country's farms. Herbicides and insecticides are also used extensively in the nation's forests. Thirty-five thousand insecticide formulations with more than 1,400 active ingredients are now in use. The EPA has sufficient data on the health effects of only about 10 percent of these. It is estimated that 1.6 billion pounds of pesticides are manufactured in the United States each year.

Most agricultural chemicals are members of one of two broad classes: *chlorinated hydrocarbons* and *organic phosphates*. DDT, dieldrin, chlordane, toxaphene, lindane, endrin, aldrin, kepone, and PCBs are chlorinated hydrocarbons, all of which are federally banned, restricted or at least under review. All are potent and long-lasting, and have been linked with cancer, birth defects, neurological disease, and wildlife and environmental damage.

The widely used *organic phosphates*, although they break down more quickly, are actually more toxic to higher life forms than are the chlorinated hydrocarbons. Malathion, parathion, leptophos, and the flame-retardant tris are among the better known organic phosphates. The immediate effects of this family of poisons on humans include drowsiness, confusion, anxiety, headaches, nausea, sweating, and difficulty in breathing. The symptoms are common to so many disorders that poisoning is frequently misdiagnosed. Acute effects include paralysis, convulsions, long-term brain damage, coma, and death.

The majority of the agricultural chemicals now on the market were registered before 1972, when the regulations now in effect were adopted. The Federal Insecticide, Fungicide, and Rodenticide Act of 1972 (FIFRA), which replaced a law with far more lenient standards, calls for testing and re-certification of the 600 active ingredients of agricultural chemicals that already were in use in 1972. Sixteen of the chemicals have been re-certified. A May, 1987, National Academy of Sciences report criticized the EPA for making it more difficult to introduce new pesticides while stalling on the re-registration of the more-damaging old favorites.

The re-registration process, as it is being administered by the EPA, according to the report, is slowing the introduction of possibly safer alternatives to the poisons discussed here. Chemicals are routinely used long after evidence of the threat they pose has become available. And even when the EPA eventually does "ban" a material because of its effects on human health, exceptions are routinely given to users who say they can't get by without the chemical.

Fertilizers Grow More Than Crops
Fertilizers are also responsible for the proliferation of widespread water pollution. Water in 28 percent of Kansas wells contains more nitrates from fertilizers than are allowed under federal guidelines. Water in fifty Iowa communities exceeds the standard. Groundwater across the entire farm belt is contaminated with nitrates and pesticides. Nitrates also get into the water from sewage plants, septic tanks, and industry, also affecting urban areas. Of the 4,300 private wells tested in the Northeast in a recent five-year period, 700 contained nitrates at levels higher than federal guidelines.

Excess nitrates do more than encourage growth that clogs lakes and streams. They also threaten human health. A bacteria that converts *nitrates* to *nitrites* thrives in the alkaline stomachs of infants, especially those under four months old. The nitrites can render the hemoglobin of the infant's blood incapable of carrying oxygen; this condition, known as methemoglobenemia, the "blue-baby syndrome," results from the shortage of oxygen. It can be fatal. Nitrates themselves may be carcinogenic and in adults' digestive tracts they produce nitrosamines, proven carcinogens. One recent study showed that people living in Iowa, Nebraska, and Illinois, areas with excess nitrates in the water, have higher rates of leukemia, lymphoma, and other cancers. Many other areas have similarly elevated nitrate levels.

Salt and Our Water Supply

As irrigation water passes through the soil, it picks up salts, especially in arid climates. Rivers into which this salty water flows become brackish. Evaporation further concentrates the salts.

When water containing high levels of salts is used for irrigation, crystals build up around plants' roots, making it increasingly difficult for them to take up the water. Plant stress, reduced productivity, or death results. Millions of acres of farmlands are seriously threatened by salt pollution. The Colorado River is six times saltier today than it was before the advent of modern irrigation systems.

Increasing salinity also causes problems for cities. Removing the tons of salts that may pass into a supply system each year is an expensive and, so far, seldom-attempted feat. But high salt levels are posing a threat to water users with hypertension (high blood pressure). Salts can also react with other impurities in the water to create even more harmful substances.

Sewage Plants and Landfills

The 228 million tons of municipal solid waste generated each year in the United States take up more than eighteen billion cubic feet. It would make a steaming heap nearly eighty miles high, if stacked on one acre. (If you think Islip, New York, had trouble getting rid of its single barge of municipal waste in 1987, how would you like to try to peddle this one?) Cities are rapidly losing their ability to cope with such mountains of refuse.

Treatment plants often add pollutants to the water at the same time it is being "purified." Chlorine, for example, can react with organic matter in the water to form *trihalomethanes* (THMs), a family of carcinogens. A 1975 survey of eighty cities' water by the EPA initially alerted officials to the problem. One THM, chloroform, was found in *all* the samples. Three other THMs were also found in most of the cities' water. In 1980, a study showed that cancer rates shockingly were higher in cities that chlorinated their water, apparently as a result of the influence of THMs. Users of chlorinated water were found to have a *53 percent* greater chance of contracting colon cancer and a *13 to 93 percent* increase in the risk of getting rectal cancer, according to a report by the President's Council on Environmental Quality.

Some cities have started using other chemicals to remove biological contaminants as a result, but all the alternative treatments also have side effects that can contaminate water. Ozone, chlorine dioxide, and chloramine are the most commonly used chemicals. Dangerous by-products can be produced by reactions of each with water-borne impurities, but generally less than is the case with chlorine.

Fluoride added to water for its cavity-preventing properties is also cause for concern in some areas. Too much fluoride can cause bone and kidney damage. The EPA is currently reviewing amounts presently considered safe.

POLLUTION FROM THE DISBTRIBUTION SYSTEM

A variety of dangerous contaminants can enter water as it makes its way from the treatment plant to your kitchen tap. Many water-supply systems, especially in the Northeast, are old and literally full of holes. As the water mains deteriorate, asbestos, lead, and other toxic metals, as well as a variety of other potentially harmful materials, may be released into the water. Inhibitors added to the water to slow down the corrosion of supply-system pipes can themselves be toxic. Holes in pipes and pipe joints may allow contaminants to enter the system during times when it is shut down for repairs.

Old, corroded metal water mains are still in use in many communities, letting good water escape to be replaced by toxic materials. Water mains made of wood are still in use in many supply systems. The wood mains are typically more than a century old. Although immune to corrosion, such wood pipes are frequently rotten and the metal bands that hold them together are often in very poor condition.

Estimates indicate that there are more than 200,000 miles of asbestos-cement pipe in U.S. water distribution systems. An estimated sixty-five million people in this country use such water systems. A 1979 survey by the EPA found 20 percent of the cities examined had more than one millon asbestos fibers in each liter of water, and 11 percent had more than ten million fibers per liter. Sources of the fibers included asbestos-cement pipe, ore processing plants, asbestos-cement roofing tiles (in Seattle), and naturally eroding serpentine rock (in the San Francisco Bay area).

Studies in Canada and California have linked the ingestion of asbestos with an increased risk of cancer in the abdominal tract, although much research remains to be done. It has been found that fiber size has more bearing on asbestos toxicity than does the number of fibers present in the water supply. No standards governing the substance have been established. Asbestos, when inhaled, is a confirmed human carcinogen.

Asbestos-cement pipe's selling points have been its low cost and resistance to corrosion. But recent research has shown that in supply systems with "aggressive" water, even cement pipe will corrode. Water is termed aggresive if it is acidic and contains little calcium chloride, a naturally occurring compound that forms a coating inside pipes and protects them from corrosion.

ACID RAIN

Rainwater (and snow) can become acidic as the result of contact with airborne pollutants—mostly sulfur dioxide from the smokestacks of power plants and smelters, and nitrous oxide from the exhaust pipes of automobiles. The resulting acid rain may be as much as forty times as acidic as normal rainwater. The dramatic effects of acid rain on the health of fish and forests are relatively well-established. But increased acidity also deteriorates water's quality. The atmospheric pollutants associated with acid rain have also been found in a recent study to be responsible for an increased rate of respiratory ailments. Rain falling on the Catskills in upstate New York is ten to twenty times as acidic as normal. The area is downwind from numerous sources of "acid," principally sulfur dioxide from coal-burning power plants in the Midwest. A recent report by the Environmental Defense Fund concludes that the high acidity in New York City's reservoirs, fed in part by water from the Catskills, may be speeding the deterioration of those reservoirs and supply pipes, leaching additional toxic materials into the water. The two huge supply tunnels that bring water from the Catskills to New York City cannot be shut down for inspection and repair. To do so would leave much of the city without water, a situation officials don't even want to think about. Yet those tunnels, built in 1917 and 1937, may not have much life left.

THE PERILS OF PLASTIC PIPE

We've already discussed the hazards of lead pipes. You may be unaware, however, that the newer plastic pipes so popular today can also cause serious problems. A demonstrated resistance to corrosion, low cost, and ease of installation and repair make plastic pipe attractive to utilities replacing old supply lines. They are also used in many homes.

In addition to vinyl chloride (a known human carcinogen), lead, cadmium, and other metals are added to plastic pipe to increase its resistance to heat. Making matters worse, a variety of toxic compounds are employed in the primers and solvent cements used to fasten the pipes together. THMs (trihalomethanes) can form when impurities in the water react with plastic pipe's components.

Research has shown that micro-doses of all these materials can enter water passing through the pipes, but too little investigation has been done to reliably assess the health risk. Plastic pipe has, however, been banned for use in home plumbing systems in California as the result of concern about the quality of drinking water. The fact that plastic pipe is permeable to gasoline and many solvents and pesticides can also mean polluted drinking water if supply-system pipes pass through contaminated ground, as many do.

POLLUTION FROM INDUSTRY AND MINING

With only an estimated 10 percent of the more than 300 million pounds of toxic wastes generated each year by American industry being disposed of safely, it's no wonder second-hand industrial chemicals make up a growing part of the water we drink. Communities with problems at least as serious as those experienced in Woburn and Love Canal are located virtually everywhere.

The quantities and toxicity of wastes found at abandoned industrial dumps can be shocking. A dump in Toone, Tennessee, has been found to contain *350,000* fifty-five-gallon drums holding more than *sixteen million gallons* of highly toxic materials disposed of by the Velsicol Chemical Company. Benzene, aldrin, endrin, dieldrin, heptachlor, and chlordane were among the toxic chemicals found at the site.

A dozen dangerous pesticides were found in the water of that community. Some wells were found to contain carcinogenic chemicals at levels 2,000 times above the suggested limit. Residents exposed to the chemicals complained of dizziness, nausea, rashes, liver disorders, and urinary tract problems. Many unusual birth defects, including a baby born with an external stomach, were reported.

Problems caused by industrial wastes can be much harder to trace than those in Toone. Residents of the Denver suburb Friendly Hills have learned over the last several years how difficult it can be to get any response to complaints about bad water when state and federal water-quality agencies, the local water utility, and the polluter insist that no problem exists.

People in the middle-class community of 7,000 formed the Friendly Hills Health Action Group because they were afraid contaminated water was responsible for the high rate of cancer, birth defects, and other diseases in their neighborhood. The group's goal was to find out whether toxic chemicals from Martin Marietta's 5,000-acre guided missile facility southwest of Denver seeped into ground and surface water that flowed into the Denver Water Board's Kassler Station, a few miles from their neighborhood. (Martin Marietta is, incidentally, the largest industrial employer in Colorado.)

Concern about the quality of the drinking water in Friendly Hills first surfaced at neighborhood Tupperware parties in the early 1980s. Mothers of sick or dying children compared notes at the parties, and soon began to think there was an abnormally high incidence of serious disease in their community. Both the state health department and the EPA declined to conduct an epidemiological survey, despite the requests of concerned parents. As a result, residents of the community decided to undertake their own informal survey. Forms that had been developed by the residents of Love Canal when they were faced with a similar dilemma were distributed door-to-door by volunteers.

The results of the survey increased the neighbors' concern. It showed that forty-nine children under age fourteen in the Friendly Hills neighborhood had cancer, birth defects, serious problems with their immune or nervous systems, or other debilitating diseases. *Eight* children in the community had died from cancer, *five* from birth defects, and *two* from disorders of the immune system. Results weren't tabulated for those over fourteen, although a surprising number of illnesses were reportedly also found among adults. Parents in the community group believe this high rate of disease is the result of using polluted water from the Kassler station.

Martin Marietta has consistently denied that any extensive contamination of water has occurred. Officials of the Colorado Health Department, the Denver Water Board, and the EPA agreed. Both the EPA and the health department conducted studies of water quality and the incidence of disease in the neighborhood after results of the informal survey were released, but they subsequently reported there was no cause for concern.

In late 1984, the EPA announced that no significantly polluted water was present on Martin Marietta's property. A week later, however, the Air Force, which operates a 300-acre missile testing facility at the complex, announced groundwater under the site was contaminated with toxic chemicals after all. The Water Board shut down the Kassler Station soon after the second announcement, claiming it was simply because the facility was old.

Some research in Denver by members of the community group showed that Martin Marietta had been fined $1.24 million, ($500,000 of which was later suspended) by the state health agency for violations of its discharge permit since 1980. The community group confronted city water officials with the records, which revealed that a variety of toxic chemicals, including exotic heavy metals, acids, and solvents, had spilled into drainages leading to the Kassler Station many times since 1957. Hexavalent chromium, an unusually toxic heavy metal, was among the materials reported to have been released. Company officials said the fines were largely the result of bookkeeping errors.

Water department representatives claimed that even if some contaminated water could somehow have found its way to the Kassler Station, the water was diluted with pure water from other sources by the time it reached customers' taps. The department told the homeowners no supply line connected the Kassler Station to the Friendly Hills neighborhood.

Members of the neighborhood action group consulted a map of the city water system to prove that a supply line did, in fact, run from the plant to their community. The group paid for tests of water samples taken from hot water heaters in homes that were vacated before the supply station was shut down, and traces of volatile organic chemicals turned up.

In 1985, the EPA discovered polluted groundwater spreading from Martin Marietta toward the Kassler Station. Water moves slowly when confined in an aquifer—from as little as a few inches per year to as much as a few feet per day. A pollution source leaking contaminants into moving groundwater over a period of time will create a pollution *plume*—a zone of contaminated water—

in the aquifer. The contaminated zone will stretch downstream, in the direction of the groundwater's flow. The duration of the leakage and the speed of waterflow through the aquifer determine the shape and size of the plume of pollution.

The EPA team measured trichloroethylene (TCE) in concentrations of up to *300,000 parts per billion* in a pollution plume underlying the Martin Marietta property (the proposed maximum for the potent carcinogen is only 5 ppb). However EPA officials claimed the pollution hadn't yet reached the station's water intakes.

A lawsuit filed by parents from Friendly Hills and the National Campaign Against Toxic Waste is pending against the Denver Water Board, Martin Marietta, and possibly the state health department and the EPA. The complainants are reportedly asking for a total of $8 million in damages. They claim that negligence on the part of those named in the suit is responsible for the deaths and illnesses suffered by their children.

Although the suit is widely supported in the immediate neighborhood, some residents say the action group's members are simply "troublemakers," and that all they have accomplished is to drive area property values down. Representatives of the group counter that they were forced to take the action to protect their families. They say that without the lawsuits, neither the Denver Water Board nor the state and federal officials charged with protecting the quality of public water supplies would even *consider* the validity of their concerns by investigating the cause of their children's diseases.

BANNING TOXIC CHEMICALS? "KIND OF"

Two major sources of bad water are bad federal regulation of polluters, and weak enforcement of water quality laws. Under the Reagan Administration, the Environmental Protection Agency was *never* the active advocate of environmental quality and public health called for in the legislation that created the Agency.

The EPA's handling of the ban on the insecticides chlordane and heptachlor is typical of its laidback approach to the regulation of toxic chemicals. Use of the chemicals was banned in 1975, when evidence of their extreme carcinogenity was confirmed. Chlordane had been the country's leading household insecticide up to that time. But the EPA made "exceptions" to the extent that millions of pounds of these chemicals were produced and sold even after the "ban."

Velsicol Corporation, the sole manufacturer of the two chemicals (and, incidentally, owner of the waste dump in Toone), has long claimed that although the pesticides are highly toxic, they pose no risk to human health if applied in accordance with the company's recommendations. For termite control, this means placing the poisons in holes in the ground under the house. However, high levels of the compounds kept showing up in homes, according to their owners, long after they had been treated. So the company decided to prove its claims by treating some homes "according to the book," then monitoring them itself to prove no contamination of the living space had resulted.

The tests backfired. Not only were chlordane and heptachlor getting inside the homes being treated, but readings taken in some houses a year after treatment were up to *four times* those taken just after application of the chemicals. With a long history of tests associating the pesticides with cancer and damage to the immune and nervous systems, the EPA was finally left with no choice but to end its exemption to the ban. It did not do this, however, until mid 1987—a dozen years after the initial ban and twenty-five years after Rachel Carson warned of the extreme dangers of chlordane and heptachlor.

These persistent chemicals, produced by just *one* company, are now found in drinking water in most parts of the country and in the tissue of *95 percent* of all Americans. And there are literally thousands of businesses—in mining and agriculture as well as in industry—that pose an equally serious threat to water quality. Even though the protection of water quality is getting more emphasis these days, the scale of the pollution problem ensures it won't go away for decades. Luckily, those who are concerned about the quality of their water supply need not wait for others to clean up the water. An affordable and safe water supply is yours for the asking. All that is required is action: *yours*. Use chapters 8 and 11 of this book as your guide to better water.

PROPOSITION 65

The public's disenchantment with the poor state of our water supply reflected in the three-to-one margin by which California's Proposition 65, the measure calling for an end to the disposal of toxic materials in the state's waterways, passed in 1986. The strong message to state officials seems to be: "Human health comes before corporate convenience." However, Governor Deukmejian, who opposed the measure when it was a ballot issue, took the advice of the Chamber of Commerce, the chemical industry, and farmers in adopting a go-slow approach to enforcing the measure.

The Deukmejian administration included only the forty-nine toxics for which human carcinogenity is *proven* on the initial list of pollutants to be banned, spurring a lawsuit from the measure's backers. Supporters of the initiative say 250 chemicals shown to be toxic and/or carcinogenic in animals or humans were intended to be included in the ban.

The controversy over the enforcement of Proposition 65 demonstrates the difficulty of ridding ourselves of bad water. All the many states where measures increasing the protection of water quality are being considered or have recently been adopted have been the scene of intense struggles. As a resident of one of these states, you must strongly support tighter regulation of water quality and be willing to *pay* for it.

Comprehensive management and protection of water resources is time-consuming and costly. But the growing liability of being responsible for pollution and the growing public outcry for safe drinking water are improving the outlook for good water. Still, it is clear that until public policy catches up with individual concern, *each* of us will have to do what we can to safeguard our own water supplies.

4 · Bad Water and Your Health

TOXIC QUESTIONS ABOUT OUR DRINKING WATER

A virtual river of toxic substances is pouring into the nation's drinking water. And there is indisputable evidence that bad water has caused significant health problems for some of those consuming it. But just how prevalent is bad health caused by poor water quality? Is bad water causing the epidemic of ills described by some health experts?

There is still no consensus about the answers to these questions, but the evidence that equates bad water with bad health is mounting. Modern water-testing methods can detect some toxic materials at concentrations of a few parts per *trillion*. Medical researchers are now able to demonstrate that even such minute doses of some poisons can cause *severe* disorders in laboratory animals.

Some medical experts persist in saying that even the most poisonous substances, such as dioxin, pose little danger to human health when found in water. Yet others produce persuasive evidence of ruined health in individuals—and even in entire communities—with polluted drinking water. Water users are too often confronted with the distressing problem of deciding which "expert" to believe, as were the residents of the Friendly Hills neighborhood discussed in chapter 3.

This chapter will reveal the reasons for the wide variation of opinions about the health effects of water contaminants, and describe the often-tragic consequences of these differences of opinion. This chapter will also alert you to the dangers of the principal classes of contaminants found in drinking water, and describe the symptoms of exposure. It will explain how these toxics enter the body. We will explore some of the basics of the science of toxicology, and examine some possible answers to the difficult questions posed by the toxic materials found in drinking water.

DBCP—AN EXERCISE IN STERILITY

The development, use, and eventual outlawing of the pesticide dibromochloropropane (DBCP) illustrate the difficulty of keeping even an obviously potent poison from causing damage to human health, even when medical evidence of its toxicity is clear from the beginning. Despite the safety net built into the regulations designed to protect the public from toxics in drinking water, the unhappy truth is that this net too often consists of little more than theory; it is more holes than net. The story of DBCP unfortunately is far from unique. Here's how this deadly chemical, like so many others, has found its way into our drinking water and what it has done to human health.

DBCP was developed in the early 1950s by Shell Chemical Company and Dow Chemical for killing nematodes, the microscopic worms that attack plants' roots. It was designed to be injected into the soil or mixed with irrigation water. The nematocide degrades slowly after its introduction into the soil so that fields need be treated only once every three years. Growers of cotton, pineapples, citrus fruits, and other crops found their farms' productivity greatly increased after application of this soil fumigant. Farmers especially appreciated the fact that DBCP could be applied while crops were present in their fields. Previous nematocides had been so toxic that they killed any plants growing in the treated soil.

During the 1950s, toxicologists working for Shell and Dow tested DBCP's toxicity on animals. Although the testing was superficial by today's standards, the companies voluntarily put the chemical through screening typical of a period in which there was little government control over testing.

Researchers exposed rats, guinea pigs, rabbits, and monkeys to DBCP, then observed the animals for up to six months, looking for changes in weight or blood chemistry. Autopsies were then performed with special attention given to the animals' lungs, livers, and kidneys, the organs thought most likely to register damage. *The product was put on the market while the toxicity testing was still being performed.*

By 1958 the first reports on DBCP's health effects were completed. Researchers found that the organs of animals receiving 5 parts per million (ppm) of the chemical showed some damage, that their growth was retarded, and that males' testes were undersized. Male animals fed 20 ppm DBCP were completely sterile. All test animals receiving 40 ppm soon died. The researchers found no effect when the chemical was brushed across the shaved backs of rabbits, although they did note DBCP quickly passed through the animals' skin into the bloodstream.

Because of DBCP's toxicity, Dow literature encouraged farmers to wear gloves, shoes, and clothing resistant to the chemical, and to avoid breathing its fumes. Warnings on Shell's product were less emphatic. Demand for DBCP grew phenomenally during the 1960s. Georgia peach growers and California grape growers enjoyed especially dramatic increases in productivity. DBCP was also used extensively on fields growing figs, bananas, strawberries, and lettuce. Yields up to three times those before treatment were reported. By the early 1960s, millions of pounds of DBCP were being produced, and the warnings and precautions were largely forgotten.

In 1964, Occidental Chemical Corporation, a subsidiary of Occidental Petroleum Corporation, started producing a variety of nematocides containing DBCP in a plant it had recently purchased in Lathrop, California, in the fertile San Joaquin Valley. The company was undergoing a period of unprecedented growth based primarily on agricultural chemicals. The new plant in Lathrop and the acquisition of Hooker Chemical, the villain of the Love Canal disaster, were part of Occidental's rapid growth. Of the more than 200 pesticides and fertilizers produced at the Lathrop plant, about thirty contained DBCP.

The observance of safety precautions on the DBCP line at the Occidental plant were lax. No ventilation was provided, and employees frequently worked barehanded in clothing saturated with the product. None of the workers considered DBCP to be dangerous.

The only "poisons" in the Lathrop plant were thought to be the organophosphates, insecticides such as malathion and parathion, which caused dramatic symptoms when they contacted the skin or were inhaled. Workers accidentally exposed to the pesticides quickly experienced nausea, dizziness, and uncontrollable spasms of the eyelids and tongue. Atropin, an antidote for organophosphate poisoning, was administered to alleviate the symptoms. It would take only several drops of any of the organophosphates to kill a large man. DBCP seemed mild by comparison, and was considered safe by the chemical plant workers.

Occidental bought its supply of DBCP from Shell and later from Dow, diluted the chemical, then added it to a variety of products. Neither supplier warned Occidental of DBCP's apparent toxicity. Shell and Dow together produced up to thirty million pounds of the chemical annually at the time. Nematocides containing up to three million pounds of DBCP were mixed each year at the Lathrop chemical factory, largely for use in California. (One-fifth of the total of more than $1 billion worth of pesticides annually produced are used in California. Pests seem to be especially fond of the warm, moist environment provided by irrigated agriculture in a warm region.)

In 1973, the National Cancer Institute announced the results of tests revealing that DBCP and a related nematocide, ethylene dibromide (EDB), caused stomach and breast cancers in rats and mice. The researchers who conducted the experiments were alarmed because the laboratory animals developed tumors more quickly than ever before observed. The Institute notified Shell and Dow of its findings, warning the companies that additional protection should be provided for those working with the substances. Neither the companies' employees nor Occidental were notified of the findings.

By 1976, only after repeated tests had confirmed the National Cancer Institute's findings, the Environmental Protection Agency began its own belated investigation of DBCP. When the EPA put DBCP on a list of chemicals that it warned it *might* ban, Shell and Dow finally told their employees the chemical had been found to be a possible carcinogen in humans. The findings were downplayed because the companies mistrusted the test methods used (massive doses were fed directly into animals' stomachs). Occidental and other companies buying DBCP from Dow and Shell still weren't advised of the EPA's concerns.

Occidental officials realized by 1976, however, that there were health problems at the Lathrop plant. One employee who found he had become sterile during the three years he worked on the agricultural chemicals line quit and filed a workman's compensation claim after doctors advised him his sterility was work-related. Several other men who had worked on the DBCP line decided to get tested too, and found they were sterile or had very low sperm counts. During 1977, the company had all of its employees tested, and found that fourteen of the twenty-five men who had worked with DBCP products were sterile or nearly so. The longer the men had worked with the nematocide, the more likely they were to be sterile.

News of the plight of Occidental's Lathrop employees became public by the summer of 1977. Headlines in newspapers across the country and regular reports by the nation's wire services and major television networks kept the developing situation at Lathrop in the public eye. Later, it was revealed that fifty of the eighty-six employees on Dow's DBCP production line in Magnolia, Arkansas, were also either sterile or exhibited very low sperm counts. This greatly increased concern, since Dow was supposed to know more about the chemical than anyone and had a reputation (although undeserved) of adhering to strict safety precautions in its chemical factories.

Media coverage of the difficulties caused by DBCP resulted in growing public alarm. The publicity spurred a flurry of activities by the state and federal agencies responsible for regulating dangerous chemicals. The California Occupational Health and Safety Agency ordered production of nematocides containing DBCP stopped until further review was completed, and the California Department of Agriculture temporarily suspended use of the chemical in the state.

During September, 1977, the federal Occupational Safety and Health Agency announced an emergency standard of 10 parts per billion for exposure to DBCP in the workplace. The limit was later lowered to 1 ppb, *1,000 times* lower than Dow's long-standing recommendation. The EPA announced a suspension of the chemical's registration for use on squash, lettuce, tomatoes, and several root crops, and the federal Food and Drug Administration announced an investigation of DBCP residues in food crops during September, 1977. Growers of perennial crops such as peaches, pineapples, citrus fruits, grapes, and nuts were allowed to continue using the nematocide.

Then, in 1979, it was discovered that hundreds of wells in the San Joaquin Valley were contaminated with DBCP. Part of the contamination had come from Occidental's plant in Lathrop. A 1975 memo showed that company officials were aware even then that the wastes dumped for years behind the plant were contaminating wells on the company's property and nearby. The memo stated that an estimated 1 to 3 percent of the pesticide factory's product ended up in the unlined waste lagoon out back—an estimated five tons of pesticides annually diluted in 10,000 tons of waste water. The company officer who wrote the memo suggested a quick halt to dumping as the most prudent course. He was concerned that state water-quality officials would find out about the contamination. His suggestion was ignored.

As it turned out, the DBCP pollution in the valley was much more extensive than could have been caused just by the leakage from the Lathrop dump, significant though that was. It has now been determined that groundwater underlying *7,000 square miles* of the valley contains measurable concentrations of DBCP. Use of the product on the area's farms was the only way such extensive aquifer contamination could have occurred.

Until the valley pollution was discovered, it was believed that

DBCP was filtered out of water as it passed through the soil, protecting underlying aquifers. DBCP's presence in the region's groundwater and its subsequent discovery along with other pesticides in aquifers underlying agricultural communities across the country disproved that theory. Eventually estimates indicated that fully *one-half to one-third* of the groundwater underlying California's major agricultural areas were contaminated with DBCP at concentrations of up to 5 parts per billion. Wells containing more than 1 ppb were closed. Other areas also suffered DBCP pollution. Concentrations of *more than 18 ppb* showed up in some Arizona wells. Traces of product were also found in community water systems in Los Angeles and Hawaii.

In light of continued confirmations of the nematocide's extreme carcinogenity in animals and findings that it could also cause chromosomal damage, the EPA was forced to institute an almost-complete ban on the product in late 1979. But pressure for continued use of DBCP was intense, even in the midst of the publicity about its toxicity. Representatives of the Federal Department of Agriculture, the State of Hawaii, and farmer's groups testified in EPA cancellation hearings that financial ruin would follow a ban on the nematocide. Hawaiian pineapple growers were granted the only exemption to the ban, but a thriving black market for DBCP developed in other parts of the country after the ban took effect—with California grape growers leading the list of illegal customers.

DBCP'S FALLOUT

The DBCP debacle is important because it reveals many of the characteristic flaws in the nation's effort to protect its citizens' health from the effects of polluted water. The failure of Dow and Shell to take seriously the warnings of their own researchers and the fact that the companies didn't warn their workers unfortunately reflect practices that are far from unusual. The callous attitude of Occidental and its subsidiary, Hooker Chemical, toward the plight of its employees, its customers, and the general public throughout the DBCP debacle is equally or more disturbing.

The indifference of the officials responsible for protecting the public's health is at least partially understandable. Even the most zealous regulators are overwhelmed with a task that is orders of magnitude beyond their budget. And all government agencies involved in protecting drinking water quality are subject to the fickle winds of politics. Overeager enforcement of water quality laws can cause serious political repercussions for the agency in question, especially if the polluter turns out to be an influential company that is a cornerstone of the local (or national) economy.

A laissez-faire approach to regulation is too often the result. This was certainly the case in Lathrop. So far, regulatory agencies have amassed a dismal record in safeguarding the purity of the water we all drink. Pesticides, since they are designed to be poisonous, receive more stringent handling by these agencies than do the host of other potentially toxic materials in daily use in this country. Even then, however, DBCP was widely used for twenty years after its danger was initially noted. The chemical slipped right through the safety net of regulations and agencies designed to keep workers and the public safe from toxic substances in the workplace and in our drinking water. With such agencies taking their stewardship this lightly, it is no wonder few trust government to be the guardian of the nation's water quality.

It also becomes increasingly difficult to trust industry's persistent claim that no further enforcement is needed, that industry can be trusted to monitor itself. In the case of Hooker and Occidental, their officials showed little interest in anything other than maximizing short-term profits throughout the twenty years DBCP was in use. They even toyed with the idea of turning a profit on the sterility problem. Once that problem was made public they considered trying to patent DBCP as a male contraceptive! The DBCP line was even restarted, for a short period, using already sterile workers, with the idea that they were ideal for the job since damage to them had already been done!

During a period of suspended operations, company officials made plans to re-enter the market, expecting temporary restrictions to be lifted despite the mounting evidence of DBCP's dangerousness. Their projections showed the profits to be made by selling DBCP products could easily outweigh the potential damages paid to those bringing lawsuits. The 1979 ban on DBCP halted these plans.

Only the publicity surrounding the involuntary sterilization of the men at the Lathrop plant and the dramatic news of aquifers across the nation being poisoned with the chemical forced the EPA to ban it. Otherwise the intense pressure from the product's users would undoubtedly have resulted in a continuation of the partial ban instituted in 1977, as has happened with so many other substances at least as dangerous as DBCP. The lack of consensus on how to reliably assess a chemical's toxicity is a big part of the reason why such a distressing state of affairs exists.

THE DBCP THREAT TODAY

Contaminated water has spread the DBCP damage far beyond the boundaries of the factories where it was produced and the farms on which it was used. Valued for its persistence after introduction into the soil, the chemical is even more tenacious once it gets into an aquifer. Even though it has been banned for about a decade, groundwater polluted with DBCP can still be found *everywhere.*

During the 1980s, it was discovered that rates for childhood leukemia in California's Central Valley are highest in those areas where DBCP contamination of groundwater is worst. A 1982 survey by the California Department of Health Services found leukemia rates for those under twenty were approximately *eight times the national average* in areas where groundwater contained more than 1 part per billion DBCP. The leukemia rate for wells with moderate DBCP contamination (between .05 and 1 ppb) was about *four times average*. The rate in areas with even a low concentration of DBCP in groundwater (less than .05 ppb) had about *twice* the expected childhood cancer rate. More than 1,500 wells have been closed in California because water drawn from them contains more DBCP than the maximum concentration allowed by state law, 1 part per billion. Future studies may well reveal similar patterns in other parts of the country.

WHAT IS TOXIC?

There are *some* good reasons for the diversity of opinions about the health effects of water-borne contaminants. The sheer number of potentially harmful materials found in today's water is a leading factor. About 35,000 toxic compounds are now recognized by the EPA. Testing a material's toxicity can take several years and cost millions of dollars, which is the biggest reason reliable data are not available for most.

Add the fact that unexpected health effects can result from the *interaction* of two or more contaminants, and the water picture becomes even cloudier. Very little research has been done on the *combined* effects of multiple pollutants.

Individual sensitivity to a given toxic also varies widely, making definitive conclusions even more difficult. A concentration of a poison that will cause no noticeable effect in a healthy adult may trigger a severe reaction in an infant or in a person already in poor health.

Inadequate or varying results of toxicity testing can also delay action on substances that are eventually proven to be harmful. For instance, no consensus exists today on the effects of *asbestos* in drinking water. Although extensive testing has shown the material to be a potent carcinogen when inhaled, equally exhaustive tests have yet to be done to definitely determine its effect when swallowed in drinking water. Until *several* researchers have obtained the *same* results following the *same* test procedures, no matter how ominous their results, scientists consider tests to be inconclusive.

The early experiments that have linked the ingestion of asbestos with an increased risk of cancer of the digestive tract are not yet considered sufficient evidence to *prove* it is dangerous when swallowed. Many people, however, feel that waiting for that absolute proof may prove hazardous to their health and that of their families. More than *one-tenth* of the water from cities included in a 1979 EPA survey had more than *ten million* asbestos fibers in each liter of water. Some urban supplies far exceed even that disturbing figure.

Even when *repeated* tests show beyond any doubt that a chemical causes a certain disease in laboratory animals, a great deal of

controversy can remain, as we've seen was the case with DBCP.
The manufacturers and users of the substance in question routinely
disagree with the negative findings, usually claiming that if the
product is used "properly" no danger to human health will result.
And even if it is admitted that *some* danger may exist, the product's
defenders normally contend its usefulness outweighs any potential
risk. It can, and usually does, take decades from the time a
substance is first suspected of causing damage in humans until it is
banned.

The use of tests on animals to establish human susceptibility to a
chemical's effects is especially controversial. Testing on animals,
however, is the only reasonable alternative to allowing *any* substance to be marketed for *any* use—the only alternative to simply
assuming, in effect, that it is safe until proven otherwise. That
assumption would make guinea pigs of us all. Most independent
researchers reject this approach. They say that since virtually all
known poisons have been shown to be toxic to *both* humans *and*
animals, requiring proof of damaged human health before banning
a product is unconscionable.

THE MAJOR TOXICS AND THEIR EFFECTS

A substance is termed "toxic" if it has a negative effect on an
organism's health. *Acute* effects are those that occur soon after
exposure to a toxic material. Symptoms range from rashes, nausea,
headaches, dizziness, and lack of energy to convulsions, coma,
and death, depending on the severity of the poisoning. Such
reactions occur anywhere from a few seconds to several hours after
initial exposure. Disorders of the throat, stomach, kidneys, liver,
intestines, and lungs, and unusual susceptibility to infections can
also result from acute poisoning, although such symptoms may take
longer to manifest.

Chronic effects result from exposure to a toxic substance over a
longer period of time and take longer to show up. Although the
concentration of the poison is usually lower, the health impact of
chronic exposure can be even more serious than the effects of acute

poisoning. Cancer, disorders of the heart and circulatory system, and nervous and immune system damage are among the more short-lived effects of chronic poisoning. Birth defects caused when a toxic material disrupts the development of a fetus can have impact on an entire life; and altered chromosomes can cause damage that will impair the health of generations to come.

The chronic effects are more difficult to pin down. Deciding precisely what substance, or combination of substances, caused a serious disorder that may appear decades after initial exposure can be a challenge, to say the least. People may be exposed to countless toxics over the years. The long latency period of certain chronic effects—up to forty years for some cancers—makes prediction all the more difficult.

All that can be traced with any degree of certainty is an increase in *rates* of cancer and other diseases in a large population exposed to the same contaminant, and even this research can be quite difficult. Keeping track of such a group for a period of many years in our mobile culture is no easy task in itself, and conducting such an epidemiological study is enormously expensive.

Money to pay for epidemiological investigations of the chronic effects of contaminants in drinking water has only been available for about the last decade, and then often at insufficient levels. Confirmation of the experimenter's findings by others can take another decade or so, enough time for more of those exposed to become statistics.

Chronic exposure to a toxic substance can damage the body in a variety of ways. Only about thirty substances are *known* human carcinogens, but strong evidence of human carcinogenity exists for more than 200 materials as the result of conclusive animal tests. About 10 percent of the 8,000 substances to which animals have been exposed in cancer tests have produced at least some evidence of cancer. Radioactive materials, such as uranium, radon, arsenic, benzene, and vinyl chloride are some of the known carcinogens *commonly* found in drinking water.

A *mutagen* causes changes in a cell's genetic material that is inheritable from cell to cell and sometimes from generation to generation. Most known and suspected human carcinogens are also mutagens. Mutations occur in our cells constantly, usually causing no effect on our overall health. In some cases, however, a cell will continue to grow and divide after a mutagen has altered its basic genetic structure, often resulting in disease. If chromosomes in a sperm or egg are altered, birth defects, inherited diseases, disorders of the nervous and immune system or the brain, and other abnormalities may be passed on to our offspring.

Pesticides such as the herbicide 2,4-D and the insecticide hepatachlor, industrial wastes such as dioxin and PCBs, the solvent trichloroethylene, and most radioactive materials are mutagens that are often found in drinking water.

A *teratogen*, which means literally "monster causing," damages the developing embryo and fetus, resulting in birth defects. Most carcinogens can also have teratogenic effects. Lead, the solvent carbon tetrachloride, and the trihalomethane chloroform are a few of the better known teratogens found in much of our drinking water.

A toxic material can cause a variety of other damaging chronic effects once inside the body, including sterility, chromosomal damage, miscarriage, damage to the nervous and immune systems, and disorders of the heart and circulatory system. Some researchers believe all carcinogens are capable of producing birth defects (both by directly influencing the development of the fetus and by causing chromosomal damage) when given the opportunity to do such damage, although this hasn't been proven.

Once on the loose in the human bloodstream, the kind and severity of damage a toxic material can inflict varies. The age and general level of health of the individual in question and the balance of minerals and vitamins present in her/his body will help determine the harmful effects. There are, however, characteristic kinds of health damage caused by members of the broad classes of contaminants that are found in drinking water.

HEAVY METALS

Heavy metals, for instance, are most likely to cause birth defects and developmental problems in children, and to attack the central nervous system and the brain. A recent study in Boston concluded that *nearly half* of the birth defects suffered in that city were associated with lead! Other research has shown that cadmium, aluminum, and mercury can have similar effects. The introduction of unleaded gasoline was a response to such findings, but getting the lead out may turn out to be only a partial solution.

The metal manganese, which is used as a substitute for lead in gasoline, has itself been found to be associated with degeneration of the central nervous system. Although it is an essential trace nutrient, manganese is toxic in higher doses. Symptoms of manganese poisoning resemble those of Parkinson's disease: jerky muscle movement, slurred speech, sleepiness, leg cramps, a spastic gait, emotional distress, and a fixed, mask-like expression of the face.

A survey in Newark, New Jersey, showed that the level of both lead *and manganese* in children was greater *the closer the child lived to a heavily traveled street*. While the lead and manganese absorbed by the children in the Newark study were thought to be the result of air pollution, both metals commonly find their way from streets into water supplies.

PESTICIDES

It is harder to summarize the effects of *pesticides*. With nearly 50,000 products on the market designed to kill weeds, insects, rodents, nematodes, fungi, and countless other pests, it's no wonder the human health effects are so diverse.

Nevertheless some generalizations can be made. The *chlorinated hydrocarbons* (aldrin, dieldrin, endrin, chlordane, heptachlor, DDT, toxaphene, and kepone are some of the best known) are the most dangerous because of their persistence and their resulting ability to build up in living tissue. A toxic dose can be accumulated over the years even if the day-to-day intake is below the threshold of toxicity. Meat eaters at the top of the food chain (such as humans) suffer the most, since any tainted flesh they consume will pass its contaminants on to them. As a result, use of chlorinated hydrocarbons has been strictly controlled or banned. But because of their powerful persistence, chlorinated hydrocarbons are still found in water virtually everywhere. The highest concentrations in this country are in the lower Mississippi basin.

Irreversible damage to the central nervous system is the most obvious effect caused by many of the chlorinated hydrocarbons in humans. In addition, most have been associated with cancer in tests on animals, especially liver cancer. Typical symptoms of exposure result from the chemicals' effects on the nervous system. They include headaches, blurred vision, dizziness, involuntary movement of the muscles, and personality disorders.

The still widely used organophosphate pesticides, which include malathion, parathion, leptophos, and more than one hundred other compounds, are toxic because they interfere with the function of the nervous system. They constitute a lesser (but still substantial) threat to human health than chlorinated hydrocarbons because they break down more quickly once introduced into the environment. Organophosphates range in toxicity from the relatively innocuous malathion to the extremely toxic parathion. Symptoms of mild acute exposure resemble those of the flu: headache, blurred vision, nausea, weakness, and abdominal cramps. More severe exposure is characterized by loss of coordination, constricted pupils, tremors, muscle twitches, and a tight feeling in the chest that may be mistaken for a heart attack.

Delayed neurotoxicity, a degenerative disease of the nerves that resembles multiple sclerosis, may manifest itself after exposure to organophosphates. The victim suffers paralysis of the affected parts of the body. Few *chronic* effects have been well-documented for organophosphates because of the chemicals' relatively short lifetimes.

RADIONUCLEIDS

Radionucleids are dangerous because they are constantly emitting charged particles that damage whatever they hit. Radium, uranium compounds, and radon, a uranium decay product, are the most likely to be found in drinking water. Although much of the damage caused by radioactivity is repairable, some is not. Concentrations in drinking water are far below levels that cause acute effects, but the constant bombardment of the body's cells by radioactive materials can cause mutations in cells, some of which can be passed on in the form of birth defects. Radiation is also a confirmed cause of cancer in humans and animals. Polluted drinking water is a primary source of indoor air pollution by radon.

PETROLEUM PRODUCTS

Petroleum products are found virtually everywhere. Leaking underground storage tanks and accidental spills on the surface are the main ways such products as gasoline, diesel fuel, kerosene, and the countless other petrochemical fuels and lubricants find their way into drinking water. Benzene, lead, manganese, toluene, xylene, and ethylene dibromide (commonly called EDB and used as a nematocide until banned in 1984 for most uses) are just a few of the dozens of toxic substances that are to be found in such products.

Because of the diversity of dangerous components of gasoline and other petroleum products, it is hard to generalize about their health effects. Kidney problems and dry, irritated skin can be among the first signs of contamination. The distinctive "filling-station" aroma of benzene, which can be detected by the human nose in concentrations as low as a couple parts per billion, is a sure sign there is a problem. If you are one of the many who thinks your water sometimes smells like gas, your thinking is correct. Cancer, birth defects, nervous-system damage, and a host of other chronic effects can be caused by gasoline's components.

INDUSTRIAL WASTES

Industrial wastes also come in so many forms that it is difficult to generalize about their health effects. They can include any of the substances just described as well as a chemical cornucopia of other toxic substances. *Solvents* are probably the most commonly found in drinking waster. Trichloroethylene (TCE) is one of the most widespread. It has been found in groundwater in every part of the nation. TCE acts as a *depressant* of the central nervous system, and has been shown to be a carcinogen in animals.

Polychlorinated biphenols (PCBs) are a family of about sixty chemicals used primarily in the electronics industry. They are extremely persistent, once they have entered the environment. The Hudson River in New York and parts of the Great Lakes have suffered some of the most severe PCB contamination in the nation, although the chemicals have been found in all parts of the country. PCBs are especially toxic to the liver and are a probable cause of nervous system damage. They can cause a skin disorder called chloracne in humans, which is like a permanent case of juvenile pimples. PCBs have been found to be carcinogenic and mutagenic in animals.

Dioxin is an impurity found in a variety of products including the herbicide 2,4,5-T and the wood preservative pentachlorophenol (PCP). This *extremely* toxic chemical has been found to cause cancer and mutations in animals and skin disorders in humans.

TOXIC SHOWERS IN YOUR HOME

Our body has built-in defenses against toxic materials, but these defenses are now frequently being overwhelmed. Skin is our first line of defense against irritants in the environment. About three-quarters of the chemical substances we are likely to come into contact with are repelled by it. Among the other quarter, however, are some extremely toxic compounds, many of which are commonly found in drinking water. Solutions that are strongly acidic or alkaline, solvents such as trichloroethylene, and *most detergents* can find their way directly into the human bloodstream through the skin. Many pesticides and gasoline components are also among the multitude of dangerous substances that can be absorbed directly into the body through the skin.

Pores and hair follicles on the skin's surface provide the easiest routes of entry. Sunburn, scrapes, and cuts can greatly increase the number of substances that can penetrate the skin and the speed with which they do it.

To make matters worse, showering and, to a lesser extent, tub bathing, can have a decidedly deleterious effect on indoor air quality, the air you *breathe*. Many toxic substances will evaporate readily when exposed to air, a process called "aeration." When water contaminated with such elements (known as "volatiles") is sprayed from a shower head or run from a faucet, they will escape into the air. Dangerous concentrations of toxics can build up in the home atmosphere as a result.

The level of contamination in the air in a bathroom while the shower is in use can be especially high. Radon, chloroform, and other trihalomethanes, benzene and other gasoline components, and many pesticides are among the volatile toxics you may be inhaling in your bathroom, sauna, or spa room. Recent articles warning about the health hazards of frequent hot-tubbing and sauna use have neglected to mention these additional toxic-related hazards. Swimming pools can also pose toxic problems.

Aeration is prescribed by the EPA as the most effective means of removing volatiles from community drinking water systems. The water is cascaded through a tower designed to maximize evaporation of the volatiles. A rapids or waterfall on a surface stream has a similar effect. That's why concern has recently been expressed about tourists breathing polluted mist at Niagara Falls! Many of the toxic elements that enter the Niagara River upstream from the falls are volatile compounds that vaporize as the water pours over the falls and become mist pollution. Such cleansing of a river (at the expense of air quality) is one reason volatile elements are found in far higher concentrations in aquifers than in surface water.

What all of this means is that even volatile toxics that can't penetrate the skin may enter the lungs. Although the nose and throat will filter out the larger impurities, most of these toxic volatiles will be drawn into the lungs and do damage there or will pass into the bloodstream.

This means that purification of *drinking* water alone isn't enough. If your water is tainted with solvents or one of the numerous other toxics to which the skin is nothing more than a porous membrane, the entire water supply must be cleaned up to avoid exposure to dangerous chemicals. One study concluded that in the course of a fifteen-minute shower you can absorb as many toxics into your body as you would if you drank eight glasses of the same polluted water. Those who scrupulously filter the industrial solvents out of their drinking water should give as much thought to filtering them out of their showers—so that these poisons don't enter through their skin. And those who revel in long showers or protracted hot-tubbing should give thought to some real perils.

SWALLOWED TOXICS

Drinking and cooking with polluted water introduce many toxics into the body. Some are neutralized in the stomach while others may actually be made more toxic there, as are nitrates when converted to nitrosamines in the stomach. Many of the toxics that survive passage through the stomach are absorbed in the small intestines and pass into the bloodstream.

From the intestines, the blood goes directly to the liver, where most poisons are at least partially filtered out, chemically altered to reduce their toxicity, and passed along to the kidneys for elimination. Since the stomach, the small intestines, the liver, and kidneys play the lead role in defending against harmful substances that have found their way into the body, these organs are among the most likely to incur damage from water-borne toxics. Kidney and liver troubles are often the first sign of poor drinking water quality. Such disorders can be caused by any of the toxics that enter the body.

Although the liver removes many toxic materials from our blood, some poisons pass through without being neutralized because they are similar enough to beneficial elements to escape detection. Some, such as benzene, are actually made *more* deadly by the effects of the liver's enzymes.

While it is impossible to say with any accuracy what toll toxics in drinking water have taken on human health, it is clear that the effects are substantial. It will be decades before epidemiologists will be able to summarize this tragic toll. In the meantime, the most prudent course is to avoid becoming a victim of this chemical fallout by taking action.

For detailed information on the toxic compounds mentioned in this chapter, their effects, and how to get rid of them, see chapter 6. For information on how and where to have your water tested and how to compare and buy water-purification equipment, consult chapter 8.

5 · Where Is the Bad Water?

IT'S EVERYWHERE

*P*olluted drinking water can be found anywhere. Sources of *potential* pollution, obviously, are most numerous in highly developed areas. Larger concentrations of people, cars, businesses, and industries simply generate a larger volume and variety of possible pollutants. But there is bad water wherever there are people and even in many areas where there are none. Pesticide spraying in forests, rainfall tainted by air pollution, and disease-carrying organisms can foul even the most remote and apparently pristine body of water.

In fact, there is undoubtedly more pollution, *per capita*, in *rural* areas than there is in most urban areas. The volume of sewage and garbage is much larger in town, making efficient handling a necessity. More money is available in heavily populated areas, not only for proper waste disposal, but also for monitoring water quality, enforcing pollution laws, and cleaning up contaminated water. In rural areas the extraction and processing of resources, whether those are crops, livestock, minerals, or timber, are usually the predominant means of livelihood—*and* the primary sources of polluted water. Individual septic tanks and wells further compromise rural water quality.

Illegal dumping of toxic materials is also easier in an area where few people live. As a result, much of the toxic waste that isn't hauled to an authorized disposal site (nine-tenths of the total volume, according to the EPA) ends up in rural ditches and watercourses. While rural businesses and industries are the source of some of this illegal waste, much is hauled from more developed areas. "Midnight dumpers" specialize in the discreet disposal of truckloads of toxics, for a fee. They will often unload in town or nearby, but instances of toxic wastes being illegally dumped far from the urban area where they originated are becoming increasingly common.

So the correct answer to the question "Where is the bad water?" would have to be "everywhere." This chapter will go beyond that oversimplified answer to provide details of just what constitutes "bad" water and to tell you which regions have the worst drinking water.

WHAT IS BAD WATER?

Bad water, more than anything else, is water that has adverse effects on our health. The concentration, toxicity, and variety of contaminants in water can all influence health. But just looking at the number of Superfund toxic waste sites or the number of known instances of drinking water contamination in *your* area isn't necessarily going to tell you how "bad" your water is. An area with few known water quality problems may have pure water, or it may simply have weak enforcement of water quality regulations, masking very bad water.

The quality and capacity of water sources, the condition of the pipelines and canals that bring the water to the treatment plant, the method of treatment, the condition of the water-distribution network, even the attitudes of the local water utility's management and employees will ultimately affect the healthfulness of the water that comes out of your tap. And the condition of that tap and of the plumbing system that it's connected to are also important. All these elements have the capacity to convert good water to bad water.

POLLUTED GROUNDWATER

The discovery of serious pollution of the nation's lakes and streams—surface water—during the 1950s and 1960s led to an unprecedented cleanup during the 1970s. Rivers such as the Hudson, the Potomac, and the Ohio, which had become thoroughly fouled as the result of up to a century of overuse and abuse, were restored, largely through the expenditure of billions of dollars on improved treatment of sewage and industrial wastes.

There are still significant problems with surface water, as we'll see, but the bane of the 1980s (and 1990s) is polluted *groundwater*. This pollution began long ago, of course, but few were aware of it during the time when national concern focused on surface water quality. In reality, ground and surface water are often linked. Contaminated water from lakes and streams can flow into an aquifer, especially when surface water levels are high, and on the flip side, low water in a river can increase the flow of polluted groundwater into it. It has been estimated that one-third of total U.S. streamflow comes from springs—that is, from groundwater. Only during the last decade have we recognized the vulnerability of aquifers to contamination from toxic materials filtering down through the ground and entering from polluted lakes and streams.

The fact that groundwater can't be seen and is difficult and expensive to analyze are primary reasons it took so long for its poor condition to be discovered. The long-held belief that toxic elements are filtered out of polluted water as it soaks through the sand, gravel, or fractured rock above the water-bearing strata has also contributed to this dangerous oversight.

Even now, far too little is known about the movement of water inside aquifers. Toxic materials can be carried many miles, or can stay relatively close to their source, depending on the speed of the water's flow. Toxic chemicals leached from wastes dumped into an eight-acre site that had formerly been a sand and gravel pit near Charles City, Iowa, entered groundwater and were transported *fifty miles* to Waterloo, Iowa. A veterinary medicine manufacturer had, over a period of twenty-five years, dumped more than one million cubic feet of arsenic-bearing wastes at the site. Channels in the

fractured limestone underlying the waste dump sloped to the southwest, toward Waterloo. The Cedar Aquifer, from which more than
300,000 Iowa residents get their drinking water, was contaminated
as a result.

It is impossible to predict accurately the movement of contaminants in an aquifer without extensive monitoring. One well can
be severely contaminated while a neighboring well, drawing water
from just outside the plume of pollution, can be unaffected. Contaminants such as gasoline that float on top of the groundwater may
foul water when the aquifer is full in the spring, then stick to the
materials composing the upper part of the aquifer as the water level
falls during the summer. Groundwater contamination will then stop
until rising water levels the next spring cause a new surge of water
that smells like a service station.

One of the surprises that has come with the discovery of the
contamination of the nation's aquifers is that the concentration of
pollutants is *far higher* than those found in surface water. For
instance, the 300,000 parts per billion of the solvent TCE found in
the aquifer near the Friendly Hills neighborhood (see chapter 3) is
almost *two thousand times* the high of 160 ppb that has been
measured in surface water. Another solvent, TCA, was measured at
5,440 ppb in groundwater in Maine, *more than one thousand times*
the 5.1 ppb maximum measured in surface water. Such large
concentrations of toxics can accumulate because the water in
aquifers isn't exposed to the purifying effects of sunlight and
oxygen.

For the same reason, an aquifer is much slower to recover from
pollution. Toxics that would quickly be rendered harmless in
surface water can persist for months or years once they have
entered an aquifer, which is why such high levels of contamination
accumulate in groundwater. Many contaminants can persist even
longer. One town on the British coast that had been a whaling
center in the mid-nineteenth century was recently found to have
polluted groundwater. Analysis revealed the contaminant to be
whale oil that had soaked into the ground *a century* before!

Nineteen of every twenty gallons of fresh water in the United
States is groundwater. Aquifers provide half the drinking water and

40 percent of the irrigation water used in this country, three times what was pumped in the 1950s. Virtually all rural homes draw water from wells, and 75 percent of the nation's cities use at least some groundwater in their drinking water systems. These percentages are certain to increase as available surface water becomes more difficult (and more expensive) to find, so the news of contaminated aquifers is especially serious.

The extent of groundwater contamination has also come as a surprise. As with surface streams, some of the worst pollution has occurred in the more industrialized states, but no place has escaped the insidious infiltration of toxic materials into aquifers. Petroleum products, chemicals, and toxic wastes stored in underground tanks have found their way into aquifers across the country. Pesticides and fertilizers have tainted groundwater underlying most agricultural regions. The majority of the dumps for municipal garbage and industrial waste also threaten groundwater quality. Eighty percent of the Superfund toxic waste sites have either already contaminated or imminently endangered groundwater.

Gasoline may be responsible for as much as 40 percent of the nation's groundwater pollution, according to a recent study. The EPA estimates that a quarter of the nation's 2.5 million underground gasoline tanks leak.

Contamination with brine, the naturally occurring saltwater that is released when gas and oil deposits are tapped, is a problem in seventeen oil-producing states. An estimated 65,000 disposal reservoirs in the country hold brine. Illegal dumping of brine is a widespread problem. Brine contains up to 300,000 parts per million chlorides and 150,000 ppm sodium, ten times the concentrations found in sea water. It can also contain lead, radioactive strontium, and toxic hydrocarbons.

As surface disposal of toxic chemicals comes under increasing regulatory scrutiny, injection wells are becoming an increasingly popular method of disposing of toxic wastes. Wastes pumped into injection wells are introduced into formations that are a mile or more below the surface where they will, *supposedly*, be permanently isolated from adjoining supplies of fresh water. Many of the wells leak through natural cracks in rocks or through cracks around

poorly fitting well casings. In some cases, toxic wastes have actually dissolved the rocks in the formation where they were disposed
of and leaked out. Some states have outlawed injection wells.

Groundwater is being severely overdrafted—pumped out of the
ground—in many parts of the country, leaving what remains more
vulnerable than ever to contamination. Arizona, California, Idaho,
Kansas, Texas, and Oklahoma are experiencing the worst deficits.
Groundwater is also being severely depleted in parts of Arkansas,
Florida, Colorado, Nebraska, and New Mexico. An average water
level drop of three feet per year is common in these states, where
many wells are now deeper than 200 feet. Irrigation has already
been discontinued on an estimated two million acres of cropland
because the cost of pumping from such depths is prohibitive.

WHERE IS THE BAD GROUNDWATER?

THE MIDWEST

Some of the most pervasive aquifer pollution—and water overuse—
occurs in the Midwest. In Kansas, more than 25 percent of the farm
wells sampled during 1986 contained high levels of nitrates, mainly
from fertilizer and livestock wastes. In Iowa, where three-quarters
of the residents drink groundwater, a 1986 survey found that *one-
fifth* of the private wells and fifty community wells contained more
nitrates than are allowed under federal guidelines. *Half* the community wells sampled contained pesticides and other synthetic
chemicals. Contamination was most serious in corn-growing areas.
The weedkiller alachlor, marketed by Monsanto under the trade
name "Lasso," is one of the most commonly encountered. Alachlor,
a confirmed carcinogen in animals, was banned in Canada after it
was found in drinking water in southern Ontario. Massachusetts
banned alachlor in 1988, but it remains legal in most states.

Contamination with agricultural chemicals is also a problem in
many of the northern plains states, where the water tends to be
"hard" (containing a high concentration of calcium carbonate and
other minerals). In Minnesota, where two-thirds of the drinking

water comes from aquifers, water from *38 percent* of the wells tested during 1987 contained one or more pesticides. Residents of Lansing, a southeastern Minnesota farm community, had to switch to bottled water when pesticide contamination was discovered in six city wells during 1987.

A variety of toxic wastes have also contaminated groundwater in Minnesota. Some of the most serious pollution came from an eighty-acre site near St. Louis Park, a suburb of Minneapolis, where a coal tar and creosote manufacturing facility operated from 1916 through 1971. Soil below the waste dump is saturated with creosote to depths of more than sixty feet. Groundwater in an aquifer used by a *quarter of a million* people has been contaminated to a depth of 900 feet with the carcinogen benzene, pyrene, and a variety of other extremely toxic derivatives of coal tar. The plume of pollution reaches more than two miles from the waste dump. In 1980, five municipal wells and several private wells were closed because of the contamination, which promises to continue its spread since the scale of contamination makes it impossible to completely clean up.

THE NORTHEAST

Similar toxic waste sites abound across the states bordering the Great Lakes and through New England. In fact, more toxics are generated in Wisconsin, Michigan, Illinois, Ohio, New Jersey, Pennsylvania, New York, Massachusetts, and Connecticut than in Minnesota. The glacial sands and gravels that predominate in these states leave their groundwater especially vulnerable. These aquifers tend to be shallow and open to pollution. The pollution is, unfortunately, worst near the surface and near the more heavily populated areas, where it is most likely to be used for drinking water.

Ohio is second only to New Jersey in the production of toxic waste. Brine from the 40,000 active oil and gas wells in the state adds to the groundwater contamination caused by toxic waste. The brine is often disposed of in ponds designed to facilitate seepage into the ground, where aquifers are being contaminated. Illegal brine dumping occurs throughout the state and has poisoned live-stock and has tainted hundreds of wells. Five Ohio injection wells were closed in 1983 when it was found they were leaking into fresh water aquifers.

Toxics left over from New Jersey's industrial past and today's concentration of pollution-generating industry and population have given the state some of the nation's most serious problems with groundwater quality. Toxic chemicals have been found in most of the aquifers in the more developed portions of the state. New Jersey was the site of fully half the wells that were shut down nationwide in 1980 because of severe contamination.

Massachusetts and Connecticut also have especially serious problems. The long history of many small industries and a corresponding proliferation of small toxic waste sites, many of them in use for decades, are leading contributors to the contamination. A leaking underground gas tank at a service station in Provincetown, Massachusetts, caused pollution of this city's water system for almost ten years. The completion of a $4-million treatment plant in 1984, when 60 percent of the city's wells were shut down because of gasoline contamination, came just in time to save area merchants from a disastrous summer for tourism, a mainstay of the community's economy. (See chapter 3 for a description of problems caused by groundwater contamination in Woburn, Massachusetts.)

New York State has also experienced serious groundwater contamination. The most alarming contamination is in the northwest part of the state, long an area of extensive industrial activity. The Love Canal dump is the best-known but surely not the most dangerous of the toxic waste sites in the area (see chapter 1).

Groundwater under Long Island, New York, where an estimated two-and-a-half million people get their drinking water from wells, has also suffered serious contamination. Shallow aquifers and highly permeable soils make aquifers on the island especially open to contamination. The hundreds of thousands of septic tanks in use on Long Island are a primary source of pollution. The solvent TCE, used in household cleaning agents and in septic tank degreasers, is found in groundwater throughout the island. TCE also finds its way into the aquifer as the result of its extensive use in industry.

Pesticides are also found in the island's groundwater. Aldicarb, one of the most commonly found, was used as a pesticide on Long Island potato crops until it was banned by the state. Municipal dumps also contribute to the contamination. The now-closed dump at Islip was found to have generated a pollution plume 170 feet deep, a quarter-mile wide, and a mile long in the four decades it was in use. It has been estimated that half the municipal dumps in the state, like Islip's, threaten to or have already polluted the aquifers over which they are located.

Industrial waste dumps and saltwater intrusion as the result of overuse of the aquifer also contribute to the problems on Long Island. Thirty-six community water wells and numerous private wells on the island have been closed as a result of the contamination. Many more, undoubtedly, should be closed.

THE SOUTH

The southern and southeastern states are another area with serious aquifer contamination. Florida, Louisiana, and Texas have especially serious problems.

Many of Florida's aquifers are particularly vulnerable to pollution from toxic waste dumps and agricultural chemicals. Of an estimated 6,000 lagoons and ponds containing toxic wastes, state officials estimate 95 percent are not lined and 90 percent are not monitored for groundwater contamination.

The discovery in Florida wells of the pesticide ethylene dibromide (EDB), a potent animal carcinogen, during the late 1970s was one of the first warning signs of what has become a flood of bad news about supposedly good Florida groundwater. More than 1,000 northern Florida wells have now been closed as a result of EDB contamination, with concentrations averaging 6.5 parts per billion. The state's limit for EDB in drinking water is 0.1 ppb. EDB, which had been confined to Florida's citrus belt, was recently found at Deerfield Beach, on the southeast Florida coast. The city was forced to shut down one of its well fields when the EDB, which entered the aquifer as the result of a leaking underground gasoline storage tank, was detected.

Louisiana makes 25 percent of the nation's petrochemical products, and has long been a center of petroleum refining and chemical production. Widespread groundwater pollution has resulted. Between 14,000 and 20,000 waste pits containing petroleum residues and brine in the state's oil fields are a major source of groundwater pollution.

For years, Louisiana had one of the weakest toxic waste control programs in the country. The unsafe disposal of wastes from the many industries in the state and the increased transport of wastes from adjoining states for disposal in Louisiana resulted. Nearly 5,000 injection wells pump brine and toxic materials into underground formations, some more than 9,000 feet deep. White males in southern Louisiana have a cancer death rate among the top 5 percent in the nation.

Texas groundwater pollution has recently increased because of the recession in the oil industry. An estimated 5,000 of the 50,000 unplugged wells in the state are either already contaminating groundwater, or soon will. The state legislature cut $2.4 million from the fund for dealing with the backlog of unplugged wells in an attempt to balance a state budget sagging because of the decline in taxes and salaries paid by the oil industry. Texas is also one of the major producers of toxic wastes, many of which find their way into aquifers. Depletion of the Ogallala Aquifer on the Texas high plains is among the most serious overdrafts in the nation, partly because alternative sources of water in the volume required to continue current water use are far away.

THE SOUTHWEST AND SOUTHERN ROCKIES

Colorado has long been a center of mining and defense-related industry, a history that has left an impression on its groundwater. The Rocky Mountain Arsenal in Denver is the site of some of the most dangerous groundwater contamination in the country. The Arsenal was set up to produce munitions and nerve gas for World War II. After the war, some of the chemical-production facilities were leased to a private company acquired in the mid-1950s by Shell Oil. Wastes were disposed of in an unlined waste disposal area.

The first signs of aquifer contamination in the Denver area came in the early 1950s when crops irrigated with groundwater on farms three miles northwest of the Arsenal started dying. A 100-acre disposal pit lined with asphalt was constructed in 1957, but by the early 1970s contaminated groundwater was again damaging crops and livestock on farms north of the Arsenal. The pesticide DBCP, also a by-product of the manufacture of chemical-warfare agents, was among the toxic materials found in the water.

By the mid-1980s, a *thirty-square-mile* plume of contamination had spread from the Arsenal. The military spent more than $200 million between 1975 and 1983 attempting to control the spread of pollution. The Army estimates it will cost another *$5-10 billion* to clean up.

In 1986, the South Adams County Water District was forced to close its wells southwest of the Arsenal when they were found to be tainted with the solvent TCE. The Army paid $1 million for temporary water treatment and another $6 million for a permanent treatment plant for the district.

Martin Marietta's plant southwest of Denver is another area where defense business has polluted groundwater. In addition to the contamination in the Friendly Hills neighborhood (discussed in chapter 3), wastes from the facility have contaminated an aquifer used in the nearby Waterton area for drinking water.

An estimated ten million gallons of toxic wastes have been disposed of in fifty pits at the Rocky Flats weapons plant northeast of Denver. Contaminated groundwater has been found more than a mile from the site, frighteningly *within 300 feet* of the Denver Formation, a principal source of drinking water in the Denver metropolitan area. Mining in the Colorado mountains has also polluted many streams and aquifers. In Leadville, at 10,000 feet at the headwaters of the Arkansas River, high levels of lead, cadmium, and zinc have entered groundwater as the result of a century of mining.

Arizona and California have some of the most pervasive groundwater contamination in the west (and have developed, as a result, some of the strongest—but by many standards still highly inadequate—safeguards against further degradation). Only New Jersey, Illinois, and Ohio exceed the volume of hazardous waste produced by California industry. The state's large agricultural sector is also the source of much water contamination, such as that from the production and use of DBCP discussed in chapter 4. A 1986 report by the California Department of Health Services warned that pollutants in one-fifth of the state's largest wells used for drinking water contain unsafe concentrations of one or more toxic materials.

The semiconductor industry in the so-called Silicon Valley south of San Francisco has polluted parts of aquifers that provide water for 1.3 million people. Presence in water samples of the solvent TCA, a suspected human carcinogen, has forced the closure of forty private and five public wells in the San Jose area. More than $40 million has been spent in the attempt to trace and clean up the aquifer pollution there, but the control effort has so far had little effect. An estimated ninety underground tanks in the area are the sources of the contamination. Industrial solvents like TCA and TCE are the leading source of groundwater contamination in California. The worst pollution has occurred in Silicon Valley and the Los Angeles area. Saltwater intrusion into overused aquifers is a problem in many California coastal communities, as is salt buildup in water underlying irrigated croplands.

In *Arizona*, 95 percent of the groundwater pumped is used in the Phoenix and Tucson areas, where it supplies 60 percent of the drinking water. TCE contamination was first discovered in Tucson in 1981, and six municipal wells had to be closed by 1983. The pollution was traced to poor disposal practices near the city's airport by Hughes Aircraft Corporation and the Air Force. A 1987 survey concluded that the rate of congenital heart disease among offspring of mothers who were exposed to TCE before or during the first trimester of pregnancy was more than twice the national average.

Arizona, like most western states, has suffered extensive water contamination as the result of mining.

WHERE IS THE BAD SURFACE WATER?

Much more is known about the condition of surface water than of groundwater, and that condition is generally described as guarded but improving. Monitoring of water quality has been conducted for most major streams and lakes for a couple decades. More than $100 billion has been spent in the last fifteen years to reduce the volume of nutrients and toxics that flow into surface water from "point" sources to achieve this general improvement in water quality. (A point source of pollution is the outflow from a manufacturing plant, toxic waste dump, municipal sewage plant or other concentrated source of contamination.)

However, it is estimated that an even larger sum is needed just to build new or refurbished municipal sewage plants to further reduce pollution of surface water. Even in this era of improving surface water quality, many of the nation's streams contain too many dangerous materials to be used directly as drinking water. And even though fish have returned to many of the rivers they shunned twenty years ago, their flesh often contains too many toxic chemicals to be safely eaten.

With the cleanup of the more concentrated pollution sources at least well underway, *diffuse* sources of contamination are emerging as the leading threat to the quality of water, both on the surface and in aquifers. Excess nutrients, pesticides and salts in water from agricultural land, petroleum fuels and the products of their combustion (such as lead) from streets and service stations, salt water from the increasing number of roads salted during winter in cold climates, and rainwater contaminated by air pollution all degrade water quality. Diffuse pollution is harder to stop than that from point sources. Improving the efficiency of an industrial process to reduce the concentration of toxic wastes discharged into a river, for instance, is easier than changing farming methods across an entire region to reduce the concentration of salts, nitrates, and pesticides entering a river.

But even then, the elimination of the largest sources of pollution has reduced the burden of pollution being carried by U.S. streams and lakes. Out of twenty-four contaminants considered in a 1987 study of trends in water quality in streams monitored between 1974 and 1981, the average concentration of all but four of the toxic elements studied had decreased during the period. Concentrations of arsenic and cadmium increased, especially in the more industrialized portions of the Midwest, and concentrations of nitrates and chloride were up practically everywhere.

THE MISSISSIPPI RIVER

The Mississippi is the country's most polluted waterway. It drains one-eighth of North America and carries 40 percent of the United States' surface runoff into the Gulf of Mexico. Water and associated contamination from mines in Colorado and Montana to industrial facilities in Cincinnati, from widespread loading with agricultural chemicals in the Midwest to pollution from "petrochemical row" along the lower Mississippi in Louisiana, all end up in the "big muddy." More than a hundred petrochemical plants are located along the river between New Orleans and Baton Rouge.

A 1982 analysis of the river's water at New Orleans turned up 59 of the 126 toxics considered most dangerous by the EPA. It has been estimated that at peak flow as much as *twelve tons* of the solvent TCA and *ten tons* of trichlorobenzene, one of the eleven benzene compounds found in the Mississippi, pass New Orleans *each day.* Chlordane, DDT, heptachlor, aldrin, endrin, and dieldrin are routinely found in water samples drawn from the lower Mississippi, even though most uses of these pesticides have been banned for more than a decade. PCBs, dioxin, and most of the other toxic by-products of modern living are also found.

THE OHIO RIVER

From its beginning, where the Monongahela and Allegheny Rivers merge at Pittsburgh, to its mouth, where it empties into the Mississippi at Cairo, Illinois, the Ohio runs through much of the industrial heart of America. Heavy pollution of the river made it one of the first in the nation to get a serious cleanup. Efforts to improve the quality of the Ohio's water started in the mid-1940s, decades before the cleanup of most U.S. rivers began.

Although the Ohio has been greatly improved as a result of better treatment of the industrial and municipal wastes that used to contaminate it, the river still contains some of the most polluted water in the United States. With more than twenty cities and nearly 2,000 industrial concerns using the river for waste disposal, its water can normally be expected to contain traces of hundreds of chemicals. Spills and malfunctioning waste-treatment plants that introduce raw sewage and industrial waste directly into the river are also a common occurrence. Traces of *700 synthetic chemicals* were identified in Cincinnati's drinking water in a sample taken by the EPA in 1978!

The series of eighteen dams and locks constructed on the river to facilitate shipping have changed the Ohio from a free-flowing river into a series of slow-moving ponds. Tons of toxics have accumulated in the reservoirs over the years, and the river's ability to clean itself has been seriously impaired through the elimination of its fast-flowing sections.

THE EASTERN SEABOARD

Many of the rivers that flow into the Atlantic have, like the Ohio, been returned to much better condition after being little more than open sewers carrying industrial and municipal waste a few decades ago. The rivers are also like the Ohio, however, in that their water still carries many pollutants. The larger streams along the Atlantic coast, despite their universally improved condition, still have some of the worst water in the country. The headwaters of these rivers, however, are usually in fairly good shape. Areas where there is a long history of heavy industrial development have the most serious contamination. Agriculture in most of the Atlantic states contributes pesticides and nitrates to streams.

THE HUDSON RIVER

The estimated one-and-a-half million pounds of PCBs that were dumped into the Hudson at two General Electric factories from 1945 to 1976 are one of the more damaging of a host of toxic elements that have found their way into the river over the years.

Although sediments containing approximately 300,000 pounds of PCBs have been dredged from the river, it is estimated that at least twice that amount remains. Striped bass and other fish in the river contain too many PCBs to be safe to eat.

A 1987 Columbia University study concluded that if the city of New York were to add as little as 7 percent chlorinated Hudson water to its high-quality water from the headwaters of the Delaware River, as it did during the 1985 drought, PCB concentrations in the mixture would be six to seven times higher, and trihalomethanes resulting from chlorination would increase by 30 percent. In addition to PCBs, heavy metals such as arsenic, cadmium, mercury, and lead, and pesticides including DDT, lindane, chlordane, dieldrin, endrin, and heptachlor have been found in samples of the Hudson's water. Nitrate levels in the river have been increasing in recent years.

THE DELAWARE RIVER

The Delaware River supplies 13 percent of the nation's people with drinking water. About half of New York City's water supply comes from the headwaters of the Delaware in the Catskills, where it is generally of high quality. By the time it has passed Philadelphia and Trenton, however, the Delaware has picked up many of the same industrial pollutants found in the Hudson, although the concentration of PCBs is not nearly as high. Salt water moving farther than usual upstream threatens Philadelphia's water supply during dry years because of the large volume of water removed upstream for New York and Trenton drinking water.

THE POTOMAC RIVER

One billion dollars' worth of new sewage treatment facilities have reduced pollution in the Potomac by 90 percent since 1970, according to the Metropolitan Washington Council of Governments. Fish and water plants are making a comeback as a result of the improvement in water quality, and swimming is even allowed in some areas. But a century of heavy pollution has had its effects on the Potomac, as it has on most rivers in the Northeast.

Nitrogen and phosphorus from the Potomac, the Susquehanna,

the James, and the other four rivers entering Chesapeake Bay, North America's largest estuary, have upset the nutrient balance in the bay. The resulting overproduction of water plants, which die and sink to the bottom where their decay uses most of the water's oxygen, has killed oysters and other bottom dwellers, and drastically reduced the number of fish present.

THE COLORADO RIVER

Overuse and salinity are the Colorado River's biggest problems. The fact that the river's water is delivered to twelve million people living outside its basin and to millions of acres of farmland has degraded its quality. Irrigation and overdrafts have increased concentrations of salts and pesticides in the river. The Colorado is undoubtedly one of the most fought-over rivers in the world. Water departments from Denver to southern California covet it, as do farmers from Colorado's western slope to California's Imperial Valley. The Central Arizona Project, which started pumping Colorado water to the Phoenix area in 1985, left much of southern California short of water, especially San Diego, which had been "borrowing" water from the river pending completion of the CAP.

THE SAN JOAQUIN RIVER

One of the most polluted rivers in California, the San Joaquin is also one of the primary sources of water for the California Aqueduct, which delivers water to the southern part of the state. Pesticides, PCBs, and heavy metals all are found in its water. Mining, industry, and agriculture within the river's drainage area all contribute to the contamination. Water in the Aqueduct becomes even more polluted in its journey to the south; pesticides from the farmland through which it passes are the primary contaminants.

THE GREAT LAKES

Twenty-four million Americans' drinking water comes from the Great Lakes, which hold 95 percent of the surface fresh water in the country. As with the nation's streams, the quality of the Great

Lakes' water is improved from what it was twenty years ago, but is still far from good. A 1987 survey concluded that the health of Lakes Erie, Superior, Michigan, and Ontario had showed marked improvement since 1971, while the quality of water in Lake Huron had stayed about the same. Industrial and municipal waste and shoreline erosion are the principal sources of pollution in the lakes. Hundreds of synthetic chemicals, whose health effects are unknown for the majority, are typically found in samples of Great Lakes water.

A 1986 report by a joint U.S.-Canadian committee estimated *3,000 pounds* of toxics pass into Lake Ontario *each day* via the Niagara River. Four-and-a-half million Canadians' drinking water is drawn from Lake Ontario. Dioxin, PCBs, pesticides, mercury, and lead are among the toxic materials that enter the Niagara in northern New York, long an area of intense industrial activity.

Twelve million asbestos fibers per liter of drinking water were found in the water drawn from Lake Superior by the Duluth, Minnesota, water department. Tailings from an iron-ore processing plant that dumped its wastes into the lake were the source of the pollution.

Chicago has long been a leading source of pollution in Lake Michigan. In the early days of the city, drinking water was drawn from the lake and sewage was dumped into the Chicago River, which empties into the lake. Intakes for drinking water were moved farther out into the lake whenever problems with sewage con- tamination were encountered. A huge storm in 1887 caused the city's drinking water to become severely contaminated, leading to the deaths of *80,000* people from typhoid, diptheria, and cholera!

The flow of the Chicago River was reversed as a result of the disaster, sending the city's sewage effluent to the headwaters of the Illinois River and eventually to the Mississippi. The flat terrain in the Chicago area and the fact that more than 6,000 miles of the city's sewers also carry runoff from storms has led to frequent discharges of untreated sewage since that time. The surge of rainwater generated by storms frequently outstrips the capacity of sewage treatment plants. A sixty-three-mile-long network of deep tunnels is now being constructed to deal with the problem. The tunnels are designed to catch the overflow caused by storms, and channel it south of town where it will be treated before discharge into the headwaters of the Illinois River. Although the tunnel system will not be complete for several years, the Chicago River has already recovered sufficiently from the years of pollution it endured for fish to have moved back into it.

It can be stated unequivocally that every major river in the country is polluted to a serious degree, despite massive efforts in some areas to reverse the threat. And the condition of the nation's aquifers, although not fully known, is very poor in many areas. Clearly, more must be done—and soon—before the nation's sources of drinking water will be safe to use.

In this chapter, you've learned about some of the areas where it will be *hardest* to restore drinking water quality. In the next chapter, you'll find out more about some of the toxic materials commonly found in drinking water.

6 · Toxics Found in Our Drinking Water

"*P*rimitive" cultures in arid regions considered the poisoning of a well to be a very serious offense. Those found guilty of such an antisocial act often paid for their crime with their lives. The injury caused to those using water from the well and the fact that an irreplaceable resource had been rendered unfit for use both contributed to the gravity with which the crime was viewed.

Today, with hundreds of poisons routinely showing up in drinking water supplies, the contamination of wells (and lakes and streams) has, sadly, become an accepted part of doing business. Government agencies issue permits that allow companies to dispose of toxic wastes in nearby surface waters. Tens of thousands of dumps holding a variety of wastes leak poisons into underlying aquifers. Injection wells, in most states, pump toxics into the ground, where too often they end up in drinking water supplies.

Public concern about this state of affairs (and political action resulting from this concern) may help slow the *rate* at which harmful materials are entering the nation's water. But even though the pollution of drinking water is being viewed by an increasing segment of the public as being unallowable, it will be a long time before the nation's residents will enjoy reliably clean drinking water. Even if the flow of pollutants into U.S. waters could be stopped tomorrow (which will not happen), it would still take a long time to clean up the existing mess. No water source is completely safe from possible contamination.

What are these toxic substances? What are they used for? How do they get into our drinking water? What health threat do they pose? What are the symptoms of exposure?

While it isn't possible to list *all* the dangerous substances that have already been found in drinking water, you will find reviews of some of the most important and widely occurring contaminants in this chapter. The reviews are divided into the following subject areas:

- Polluted Runoff (page 96);
- Pollutants from the Treatment and Distribution System (page 198);
- Biological Water Contaminants (page 104);
- Metals Found in Our Drinking Water (page 106);
- Industrial Chemicals That Act as Pollutants (page 113);
- Pesticides Found in Drinking Water (page 121);
- Radioactive Materials (page 133);

The reviews include data on the health effects and existing or planned regulation for each substance. The following abbreviations have been used:

- *SDWA = Safe Drinking Water Act* and its 1986 amendments;
- *MCL* = the *maximum contaminant level* is the maximum concentration allowed in a community water system of a toxic material, defined as a *primary pollutant* under the terms of the SDWA. The EPA considers the cost and difficulty of monitoring for and removing a pollutant when setting the MCL;
- *MCLG* = the *maximum contaminant level goal* is the concentration of a primary pollutant that would be allowed in community drinking water if health were the only consideration;
- *ppm* = parts per million.

POLLUTED RUNOFF

The contaminants mentioned in this section enter water supplies primarily after being absorbed by rainwater or irrigation water that drains into streams, lakes or aquifers.

NITRATES, NITRITES, AND NITROSAMINES

Nitrates are found in the overflow of municipal sewage plants and septic tanks, feed lots and pastures, industrial waste water and landfills. The heaviest concentrations in the nation are in the Midwest, and result from the use of nitrate fertilizer, but contamination is found in all parts of the United States.

The conversion of nitrates to nitrites in the stomach can result in methemoglobinemia, particularly in infants (see chapter 3). Nitrites in the stomach can react with chemicals present in food to produce nitrosamines, which are suspected carcinogens in humans. Limited epidemiological studies have reported a higher incidence of cancer in communities with high levels of nitrates in their drinking water.

The SDWA requires regulation of concentrations of nitrates, nitrites, and nitrosamines in community water systems. The EPA called for monitoring to commence by June, 1988 (MCL = 10 ppm).

SELENIUM

Selenium is used in the manufacture of electronics equipment, in the production of steel, and in pigments, glass, and ceramics. Selenium is found in the waste from electrolytic copper refining and in the tailings of mines and mills associated with the production of gold, nickel, and silver.

Selenium salts are often found in soils, especially in arid areas such as California's San Joaquin Valley, where Kesterson Wildlife Refuge is located. Irrigation can leach large quantities of the selenium compounds from such soils. Although selenium is an essential trace element in animals (and humans), it is toxic in high concentrations. The disastrous effects of the selenium-contaminated water that was discharged in the Kesterson refuge on the reproductive functions of the waterfowl living there put both selenium and the refuge in the news during the mid-1980s. Water in many of the western states contains selenium.

Some plants absorb selenium from contaminated soil so efficiently that they can accumulate a dose that is toxic to livestock, as does the family of plants known as "locoweed" that grow on western rangeland. Grains can also pick up selenium from the soil; bread made from wheat grown in selenium-rich soil will contain selenium.

As noted above, selenium is an essential trace element in humans. It helps protect the body from the effects of poisoning by mercury, cadmium, silver, and thallium. High concentrations of selenium in drinking water can lead to human liver and kidney problems and to an increased incidence of dental caries. Animal tests have shown selenium to cause liver cancer and damage to the developing fetus.

The SDWA requires regulation of concentrations of selenium in community water systems. The EPA called for monitoring to commence by June, 1988 (MCL = 10 ppb).

SODIUM

Sodium gets into water from natural salt deposits, decomposing rocks, and—in areas within 50 to 100 miles of the ocean—from evaporated ocean spray taken up by rainwater and intrusion of saltwater into aquifers. The use of salts as de-icers in cold climates can add significantly to sodium levels in water. Municipal sewage effluent generally adds to the sodium concentration of surface water it is emptied into.

Sodium is essential to good health, but most people get at least ten times their body's daily needs, mostly from food. Drinking water seldom furnishes more than 10 percent of the total intake of sodium. The consumption of high levels of sodium has long been known to be a cause of high blood pressure in humans. Levels of more than twenty milligrams per liter of sodium in drinking water can be bad for the health of those who must be on low-sodium diets.

Sodium concentrations in nearly half the nation's cities are high enough so that it would be difficult to observe a strict low-sodium diet and use city water. Many bottled waters, especially mineral waters, have high concentrations of sodium.

The SDWA requires regulation of concentrations of sodium in community water systems. The EPA has tentatively called for monitoring to commence by June, 1989, although the agency has recommended sodium be removed from the list of materials that must be monitored.

SULFATE

Sulfate enters water from rain contaminated by air pollution, from the weathering of rocks, from steel mills, pulp mills, and textile plants, and from municipal sewage, which contains sulfates from household detergents. Sulfate can be found in water at concentrations between a fraction of one part per million and thousands of parts per million. Water containing more than 750 ppm can have a laxative effect in humans.

The SDWA requires regulation of concentrations of sulfates in community water systems. The EPA called for monitoring to commence by June, 1989. A secondary standard of 250 ppm is now in effect.

POLLUTANTS FROM TREATMENT AND DISTRIBUTION SYSTEMS

These toxic elements are added to community water systems for disinfection or to remove other harmful contaminants, are formed as a result of reactions between treatment chemicals and impurities in the water, or may enter the water as it passes through the distribution system.

ASBESTOS

Deteriorating asbestos-cement pipe in water-distribution systems in all parts of the United States is the most widespread source of asbestos fibers (see chapter 3). Some asbestos enters drinking water from natural sources such as serpentine rock in the San Francisco area. Building materials such as roofing tiles and siding, as well as mining and smelting operations, are also asbestos sources. The disposal of iron ore tailings in Lake Superior caused concentrations of up to twelve million fibers per liter in Duluth, Minnesota.

Asbestos is a proven carcinogen when inhaled. A study of workers exposed to asbestos revealed a higher incidence of stomach and intestinal cancer than average. Limited evidence suggests that it can cause cancer of the digestive tract when ingested in drinking water. The EPA estimates one person in a population of 100,000 exposed over their lifetime to 300,000 fibers per liter in drinking water will die of cancer.

The SDWA requires regulation of concentrations of asbestos in community water systems. The EPA called for monitoring to commence by June, 1988 (MCLG = 7.1 million fibers per liter).

CHLORAMINE

Formed by combining ammonia with chlorine, chloramines are used as disinfectants in community water systems. They were introduced as an alternative to chlorine because of concern about the formation of trihalomethanes (THMs). Chloramine is also more persistent than chlorine. Once added to drinking water, it remains longer to kill disease organisms that may be introduced in the distribution system. Fewer unpleasant tastes and odors are caused as a result of chloramine treatment than is the case with chlorine.

Chloramine can cause anemia in dialysis patients. No other effects on human health have been documented. The chemical is now unregulated but is a candidate for future regulation.

CHLORINE

Most chlorine in community drinking water systems is added as a disinfectant. The chlorination of water, introduced in the early 1900s, has virtually eliminated the spread of diseases such as typhoid fever and cholera through water systems. The fact that chlorine reacts with organic material in the water to form trihalomethanes, many of which are toxic, has caused some cities to switch to alternative treatment methods.

When introduced into water, chlorine is converted to hypochlorite ions, which do the actual work of disinfection. Chlorine and hypochlorite ions have been proposed for routine monitoring in the future by the EPA.

CHLORINE DIOXIDE

Chlorine dioxide is a foul-smelling greenish-yellow gas that decomposes with explosive force to form chlorine and oxygen. It is used for bleaching paper pulp and flour, and, increasingly, as a substitute for chlorine in the disinfection of water. It is more widely used for water treatment in Europe, but lower concentrations are used there than in this country.

Chlorite, which is produced in water treated with chlorine dioxide, has been shown to cause blood disorders in laboratory animals. Both chlorine dioxide and chlorite are candidates for monitoring in the future.

DIMETHYLFORMAMIDE (DMF)

DMF is used as a solvent in the production of synthetic textiles, in the production of paints and dyes, as a paint stripper, and as a gasoline additive. Its use as a component in glues used to connect plastic water lines is its primary source in drinking water. Animal tests on some species have shown DMF to be a carcinogen and a mutagen, and to cause miscarriage and birth defects.

FLUORIDE

The seventeenth most abundant element in the earth's crust, fluoride occurs naturally in most water. It is also added to drinking water to reduce cavities in many communities where there is little naturally occurring fluoride. About 60 percent of Americans drink water to which fluoride has been added.

Although some surveys have shown an increased risk of cancer in communities with high levels of fluoride in drinking water, there isn't enough evidence to prove this view. Some research has inferred that fluoride is a mutagen and causes damage to the central nervous system in humans, but no conclusive evidence exists. It would appear that dental fluorosis, a disorder in which teeth become mottled and brittle, is fluoride's primary threat to human health. It occurs at fluoride concentrations of more than 3 ppm. Damage to bones and kidneys can occur with still higher concentrations. Several communities have discontinued fluoridation in recent years as a result of public concern about its health effects.

The SDWA calls for the regulation of fluoride concentrations in community water systems. Monitoring has been required since June, 1987 (MCL = 1.4 to 2.4 ppm). A secondary standard of 2 ppm has been established for fluoride.

METHYLENE CHLORIDE

Methylene chloride is used in the production of paint, varnish remover, insecticides, fumigants, solvents, cleaners, pressurized spray products, fire extinguishers, and Christmas tree lights. More than 250 million tons are produced annually in the United States. The reaction of chlorine in community water supplies with impurities already in the water is a primary source of methylene chloride in drinking water. Water in nine out of ten systems surveyed by the EPA in 1975 contained methylene chloride, with Lawrence, Massachusetts, having the highest concentration at 1.6 ppb.

Methylene chloride is converted to carbon monoxide inside the body, causing changes in the concentrations of carbon-dioxide-carrying hemoglobin in the blood. Methylene chloride impairs the functioning of the central nervous system. The presence of ethanol may increase its toxicity to the liver.

The SDWA requires regulation of concentrations of methylene chloride in community water systems. The EPA called for monitoring to commence by June, 1989.

OZONE

Ozone is produced by electrical currents, and by the reaction of some chemicals with sunlight. It is increasingly being used as a substitute for chlorine in the disinfection of water. Ozone is generated on-site because it is an explosive gas. It breaks down relatively quickly once it enters the water—one of the drawbacks to its use. Another disinfectant is often added after ozonation to provide protection from harmful organisms in the distribution system. Potentially harmful epoxides are formed as a result of the ozonation of water, but little study has been done on their effects. The EPA has recommended ozone as a candidate for future monitoring.

PHTHALATES

Phthalates are widely used in the production of plastics. One of the most common, DEHP, constitutes up to 40 percent of a PVC pipe's mass. Phthalates are found in drinking water that has passed through plastic supply pipes, and in water and other fluids bottled in plastic jugs. Everyone is exposed to DEHP. Drinking water and indoor air are the primary sources of exposure.

DEHP has been found to be teratogenic in animals. Some research has indicated it may cause liver cancer in animals. Diisobutylphthalate, another phthalate, has been found to be both a mutagen and a teratogen in animal tests.

The SDWA requires regulation of concentrations of phthalates in community water systems. The EPA called for monitoring to commence by June, 1989.

TRIHALOMETHANES (THMS)

Chlorine is used in the production of many of the most toxic compounds in use today. It's little wonder, therefore, that it can react with some of the many impurities in water and with materials in the water distribution system to form toxic substances. THMs are primarily the result of a chemical reaction between organic material in the water supply and chlorine.

A 1975 EPA survey of eighty cities' water revealed that the THM chloroform was present in all systems, and three other THMs were present in most. Since chloroform and most of the other THMs found were suspected carcinogens, this finding caused concern.

A 1980 epidemiological study revealed higher cancer rates in cities that chlorinated drinking water, apparently as the result of the formation of trihalomethanes. Residents of such cities were found to have a 53 percent greater chance of contracting colon cancer and a 13 to 93 percent better chance of getting rectal cancer than the average resident of towns that didn't use chlorine.

Increased monitoring and regulation of THMs has resulted from these findings, but THMs are generally more difficult to detect than most toxics. Total THMs are monitored under the SDWA, and several of the principal THMs are candidates for future regulation. The MCL for total THMs is 100 ppb. The following are a few members of the large THM family that are commonly found in drinking water.

Bromochloromethane

Tests on animals have shown this THM to be a mutagen and to be toxic to the developing fetus. Bromochloromethane is now unregulated but is a candidate for future regulation. The EPA called for monitoring at the states' discretion.

Bromodichloromethane (Also Called Dichlorobromomethane)

Animal tests have shown bromodichloromethane to cause kidney and liver damage. Limited results have shown the chemical to be mutagenic. Bromodichloromethane is now unregulated but is a candidate for future regulation. The EPA called for monitoring in community water systems to commence by June, 1988.

Bromoform (Also Called Tribromomethane)

This trihalomethane is used in fire retardants and as a solvent. It is found in drinking water systems largely as the result of chlorination. Bromoform is a suspected mutagen and teratogen in humans. It is estimated that a lifetime exposure to 1.9 ppb is sufficient to cause one cancer death per 100,000 people. Bromoform is now unregulated but is a candidate for future regulation. The EPA called for monitoring to commence by June, 1988, for all community water systems.

Chloroform (Also Called Trichloromethane)

Chloroform is used in the production of drugs and plastics, as a refrigerant and a propellant, and as a pesticide and solvent. The primary source of chloroform in drinking water is the reaction between chlorine and humic material in the water. Chloroform is found in water in all parts of the country as a result. The following concentrations have been found in city water systems in California: Sacramento, 30–40 ppb; San Francisco, 66 ppb; San Jose, 50–70 ppb, Los Angeles, 35 ppb; San Diego, 15–60 ppb. A 1975 EPA survey of eighty cities' drinking water found chloroform in all samples.

Chloroform is a suspected human carcinogen. It also causes damage to the kidneys, liver, thyroid and immune system and hinders development of the embryo/fetus. The EPA estimates a 1-in-100,000 risk of contracting cancer from chronic exposure to a concentration of 1.9 ppb.

Chloroform is now unregulated, but is a candidate for future regulation. The EPA called for monitoring to commence by June, 1987. The U.S. standard is 100 ppb; the World Health Organization's standard is 30 ppb.

BIOLOGICAL WATER CONTAMINANTS

COLIFORM BACTERIA

Fecal coliform bacteria, although not injurious in themselves, are monitored because their presence in water indicates other, more harmful organisms may be present. They come from human and animal waste. The SDWA requires regulation of concentrations of coliform bacteria in community water systems. The EPA called for monitoring to commence by June, 1988 (MCL = a monthly average of ten per liter).

GIARDIA LAMBLIA

Giardia lamblia is a protozoan, a single-celled creature that inhabits the intestines of humans and many animals. When inside the body, the parasite attaches itself to the walls of the intestine, draws nutrients from the partially digested food passing through, and multiplies. When expelled from the body, the organism secretes a protective cyst around itself and goes into a dormant phase until another host is found. It can survive for months when expelled into cold water, although it will last only a few days in warmer water or on land, which is why the cyst is associated primarily with water drawn from cold mountain streams.

One beaver can introduce *one billion* giardia cysts per day into the water it lives in. Although beavers are often regarded as the primary source of water pollution by giardia, feces from many other animals can also be responsible, including many that don't live in the water. A Colorado survey found there were giardia cysts in all the muskrats studied, while some beavers carried none.

The SDWA requires regulation of giardia in community water systems, and the EPA called for monitoring to commence in all systems by June, 1988. Filtration with granular activated carbon is the most effective means of removing the cysts.

The parasite causes a flu-like disease sometimes referred to as the "back-packer's lament," since the cysts are often picked up by people drinking water from mountain streams. Diarrhea is the predominant symptom.

VIRUSES

Viruses are tiny particles, too small to be seen through an optic microscope, that can cause disease in man. Of the viruses that may be found in drinking water, the most threatening to human health are the enteric viruses. They are capable of surviving the effects of stomach acids and infecting parts of the intestines. Infectious hepatitis is the most serious human disease known to be transmitted by water-borne viruses. The largest known outbreak occurred in 1956 in Delhi, India. Thirty thousand people were infected when drinking water became contaminated with sewage. It is estimated, however, that only about 1 percent of hepatitis infections are transmitted by drinking water.

The SDWA requires regulation of concentrations of viruses in community water systems. The EPA called for monitoring to commence by June, 1988.

METALS FOUND IN OUR DRINKING WATER

ALUMINUM

Aluminum is the fourth most common element in the earth's crust, and is found in water everywhere. It is used in cosmetics, deodorants, and medicines, with no apparent effect on human health. Aluminum compounds are used in water treatment, but generally don't add to the aluminum content of the water.

There is some evidence that aluminum in humans causes damage to the nervous system, and brain disorders such as Alzheimer's disease. It has been associated with dementia in dialysis patients, a condition that results in death within six to seven months of its onset. Aluminum has been shown to cause epilepsy and other nerve disorders in animals.

The SDWA requires regulation of concentrations of aluminum in community water systems. The EPA has tentatively called for monitoring to commence by June, 1989, but has also proposed removing aluminum from the list of materials that must be monitored by community water systems.

ARSENIC

Arsenic is an elemental metal used in insecticides, herbicides, and manufacturing processes. Arsenic compounds are among the most widely distributed elements in the earth's crust and in the biosphere. Zinc, copper, and lead smelters emit arsenic, which often ends up in water supplies. Runoff from mines and mine tailings is frequently contaminated. Plants watered with arsenic-tainted water can pick up the metal.

Chronic exposure results in fatigue and loss of energy. Inflammation of the stomach and intestines, kidney degeneration, cirrhosis of the liver, bone-marrow degeneration, nervous-system damage, and severe dermatitis can result from exposure to higher concentrations. Epidemiological studies of people living in areas with high levels of arsenic in their drinking water have revealed an elevated rate of skin cancer, although these findings haven't been confirmed in this country.

The SDWA requires regulation of concentrations of arsenic in community water systems. The EPA called for monitoring to commence by June, 1988 (MCL = 50 ppb).

BARIUM

The metal barium is found in surface water in all parts of the country, with the highest concentrations occurring in the lower Mississippi basin. It is used in the production of other metals, paints, paper, and pesticides. It is also used in the purification of beet sugar, in the making of animal and vegetable oils, and in the electronics industry. Barium is used directly to kill rodents. Several toxic barium compounds are commonly found in drinking water.

Barium causes disorders of the nervous system and an increase in blood pressure. Strong, involuntary stimulation of all muscles results from heavier doses. A 1979 epidemiological study showed communities with higher concentrations of barium in drinking water had a higher rate of death due to cardiovascular disease.

The SDWA requires regulation of concentrations of barium in community water systems. The EPA called for monitoring to commence by June, 1988 (MCL = 1 ppm).

BERYLLIUM

Beryllium is used in the nuclear power industry, in rocket fuels, and in the production of ceramics and metal alloys. It is sometimes found in the acidic runoff from mines. Beryllium is a proven carcinogen in animals. It also caused disorders of the respiratory system, heart, liver, and spleen. It is a suspected cause of bone cancer in humans.

The SDWA requires regulation of concentrations of beryllium in community water systems. The EPA has called for monitoring to commence by June, 1989 (MCLG = 0).

BORON

Although boron doesn't occur by itself in nature, its compounds—borax, boric acid, and sodium borate—do. Borax compounds are used in the manufacture of steel alloys, glass, and pottery, in photography, and as a wood preservative.

Boron compounds can cause mild irritation of the stomach and intestines, loss of appetite, rashes and kidney disorders. It was added to the Drinking Water Priority List in 1988 by the EPA after a 1987 survey revealed that boron compounds are commonly found in drinking water.

CADMIUM

Cadmium is a bluish-white metal found in zinc, copper, and lead ores. It is used in the production of metal alloys, corrosion inhibitors, and other chemicals and in the textiles, electronics, auto, and aircraft industries. It is used as a stabilizer in PVC pipe and is present as an impurity with the zinc found in galvanized pipe. Corrosion of such pipes is a primary source of the metal in drinking water. Mines, smelters, and electroplating plants can also pollute water with cadmium.

Cadmium is a suspected cause of cancer, testicular tumors, high blood pressure, hardening of the arteries, and the inhibition of growth. It tends to accumulate in the kidneys, and kidney disorders are often the first symptom of exposure. Its irritating effect on the stomach and intestines causes symptoms similar to those associated with food poisoning: nausea and diarrhea. Cadmium has also been found to play a role in itai itai disease, which causes a softening of the bones.

The SDWA requires regulation of concentrations of cadmium in community water systems. The EPA called for monitoring to commence by June, 1988 (MCL = 10 ppb).

CHROMIUM

Chromium is used in the production and plating of metals, in photography, and in a wide variety of industrial processes. High levels of hexavalent chromium cause ulcers, respiratory disorders, and skin irritation in humans. Hexavalent chromium has also been found to affect the body's metabolism of glucose. It can be absorbed through the skin. Trivalent chromium is an essential trace mineral, and helps protect the body from the effects of vanadium. Drinking water typically provides one-quarter of the minimum daily requirement for trivalent chromium.

The SDWA requires regulation of concentrations of chromium in community water systems. The EPA called for monitoring to commence by June, 1988 (MCL = 50 ppb for hexavalent chromium, the most toxic of the chromium compounds).

COPPER

Copper is found in water in all parts of the country. It gets there both from natural and industrial sources. Much of the copper found in drinking water comes from copper plumbing systems.

Traces of copper are essential for health. Five to 10 percent of the minimum daily requirement of 2 to 3 mg is typically consumed in drinking water. Anemia, loss of hair pigment, reduced growth, and loss of arterial elasticity can result from a deficiency. At higher levels, copper is an irritant of the stomach and intestines, and can prove fatal at very high doses. No chronic effects are known in humans. Sheep are especially sensitive to copper.

Copper is regulated under the SDWA. A secondary standard of 1 ppm is based on the fact that the metal imparts a metallic taste to water that is noticeable at concentrations of from 1 to 5 ppm. The EPA called for monitoring of copper for all systems by June, 1988 (MCLG = 1.3 ppm).

LEAD

Lead is a heavy metal found in drinking water in all parts of the country. Sources of lead in streams and aquifers include mine and mill tailings, junked batteries, leaking leaded-gasoline storage tanks, and runoff from streets, which may contain lead both from auto exhaust and spilled gasoline. Lead pipes in distribution systems and lead solder in home plumbing systems are probably the leading source of the metal in drinking water.

Lead is a teratogen and can cause impotence and sterility in males and damage to the nervous system, heart, circulatory system, bones, and kidneys. It has recently been found that even relatively low concentrations of lead in drinking water can cause hearing loss and learning disabilities in children, an increased risk of hypertensive heart disease in adults, and accelerated loss of bone mass in post-menopausal women. The EPA estimates that 143,500 children suffer reduced intelligence each year as the result of lead poisoning; lead poisoning causes 118,400 cases of hypertension, seventy-five strokes, 370 heart attacks among middle-aged white males each year.

Symptoms of lead poisoning in children include hyperactivity, chronic fatigue, apathy, dullness, and insomnia. Pains in the stomach and abdomen, nausea, vomiting, and constipation are lead-poisoning symptoms that can manifest in adults or children.

The SDWA requires regulation of concentrations of lead in community water systems. The EPA called for monitoring to commence by June, 1988 (MCL = 50 ppb, though the EPA is recommending the MCL be reduced to 20 ppb).

MERCURY

One of the least abundant elements in the earth's crust, the metal mercury is used in the production of chlorine and caustic soda; in electric lights, batteries, and switches; in thermometers and barometers; in mildew-proofing paints; in the production of paper; and as an anti-fungicide treatment for seeds and plants (a use that was banned in 1976). Mercury was used in the production of nuclear weapons at the Department of Energy's Oakridge, Tennessee, nuclear facility from the early 1950s through 1966, resulting in an estimated 2.4 million pounds of the metal escaping into the environment. Groundwater under the complex is still heavily contaminated with mercury and other poisons, some of which are seeping into area streams.

The compound methyl mercury is the form of mercury of greatest concern in drinking water contamination. It causes deterioration of the human nervous system, and impairs the development of the fetal nervous system. Consuming fish from contaminated water is a more common means of infection than is drinking the polluted water, since fish concentrate mercury in their tissues. Symptoms of poisoning by methyl mercury include impairment of speech, hearing, and thought; difficulty in chewing and swallowing; and involuntary muscle movement and impaired gait.

Breathing mercury fumes can also affect the brain and nervous system. The expression "mad as a hatter" refers to the fact that hatmakers, who traditionally used mercury to soften felt, breathed the fumes and often became demented as a result. (Lewis Carroll popularized the term with his mad hatter in *Alice in Wonderland*.) This method of curing was abandoned around the end of the 19th century.

The SDWA requires regulation of concentrations of mercury in community water systems. The EPA called for monitoring to commence by June, 1988 (MCL = 2 ppb).

MOLYBDENUM

Molybdenum is used in the production of metals, ceramics, electrical equipment, chemicals, and fertilizers. Tailings of molybdenum and uranium mines and mills can leach the metal into drinking water.

Molybdenum is an essential mineral for both humans and animals. An overdose of the metal causes gout and bone disorders in humans. Cows are especially sensitive to molybdenum's effects. Cows that consume plants growing in molybdenum-contaminated soil can become severely diarrhetic and sometimes die as a result. Symptoms of molybdenum poisoning in animals also include loss of appetite, anemia, loss of hair and hair color, and bone defects.

The SDWA requires regulation of concentrations of molybdenum in community water systems. The EPA has tentatively called for monitoring to commence by June, 1989, but has also proposed removing molybdenum from the list of substances monitored in community water systems.

NICKEL

Nickel is used in the production of corrosion-resistant metal alloys and fungicides. It is an essential trace nutrient. High concentrations have caused effects on the central nervous system, disorders of the stomach and intestines, and excess blood sugar in humans. Nickel has been found to cause disorders of the heart, brain, liver, and kidneys in animals.

The SDWA requires regulation of concentrations of nickel in community water systems. The EPA has called for monitoring to commence by June, 1989.

SILVER

Silver is used in the production of metals and chemicals, in electroplating, and in the development of photographs. Silver nitrate, the most toxic of the silver salts, has been found in animal tests to cause disorders of the central nervous system and to interfere with the supply of blood to the heart.

The SDWA requires regulation of concentrations of silver in community water systems. The EPA has tentatively called for monitoring to commence by June, 1989 (MCL = 50 ppb), but has also proposed removing silver from the list of toxic materials that must be monitored by community drinking water systems.

VANADIUM

Measurable traces of the metal vanadium are found in about one-quarter of the nation's surface water, with the highest concentrations in the Southwest. Vanadium mine tailings and auto exhaust (vanadium is found in petroleum) are probably the leading sources. It tends to accumulate in humans in the liver and bones, but hasn't been demonstrated to be harmful in the concentrations found in drinking water. Vanadium has been shown to be an essential mineral for some test animals and may also be in humans, although no minimum requirements have been determined.

The SDWA requires regulation of concentrations of vanadium in community water systems. The EPA has tentatively called for monitoring to commence by June, 1989, but has also recommended removal of vanadium from the list of substances that must be monitored by community water systems.

ZINC

Zinc can get into water from decomposing rocks, from mine and mill tailings, and from fertilizers (more than 20,000 tons are annually used in fertilizers). The primary source in drinking water is corroding galvanized pipes, especially in areas with acidic water.

Zinc is an essential trace element for humans and animals. Fifteen milligrams per day are required for an adult. Inhibited growth, loss of taste, and decreased fertility can result from a deficiency. Zinc protects the body from the toxic effects of cadmium and lead.

Poisoning can result from excessive zinc in drinking water, usually the result of long-term use of water from galvanized pipes. Irritability, stiffness and pain in muscles, loss of appetite, and nausea have been reported in people consuming water containing 40 ppm zinc.

The SDWA requires regulation of concentrations of zinc in community water systems. The EPA has tentatively called for monitoring to commence by June, 1989, but has also recommended zinc be dropped from the list of materials that must be monitored in community drinking water systems. The secondary standard for zinc is 5 ppm.

INDUSTRIAL CHEMICALS THAT ACT AS POLLUTANTS

ACETALDEHYDE

Acetaldehyde is an organic chemical used in the production of chemicals. Tobacco smoke and auto exhaust both contain acetaldehyde. The highest concentrations of the chemical in drinking water have been found in Seattle and Philadelphia.

Acetaldehyde has been proven to be a mutagen and teratogen, and to have a negative effect on the embryo and fetus in animal experiments. It is unregulated.

ACRYLAMIDE

Acrylamide is a synthetic organic chemical used in the production of plastics and textiles. Seventy million pounds were produced in 1974. It is also used to treat water, and therefore may be found in drinking water systems in any part of the country.

Acrylamide causes disorders of the central nervous system in animals. The SDWA requires regulation of concentrations of acrylamide in community water systems. The EPA called for monitoring to commence by June, 1988 (MCLG = 0).

BENZENE

A hydrocarbon found in gasoline and other petroleum products, benzene is also widely used as a solvent for paints, inks, oils, plastics, paint removers, and rubber cement. It is also used to extract oil from nuts and seeds; in chemical manufacturing; and in the production of detergents, explosives, and drugs.

A suspected carcinogen, mutagen, and teratogen, benzene is responsible for blood-related disorders in humans, including anemia and leukemia. Chronic exposure causes headaches, fatigue, anemia, loss of weight and dizziness at first; then pallor, nosebleeds, and bone-marrow damage. Acute exposure causes skin irritation and drowsiness.

The SDWA requires regulation of concentrations of benzene in community water systems. The EPA has called for monitoring to commence by December 31, 1988 (MCL = 5 ppm).

BROMOBENZENE

Used in the production of chemicals and as an additive to gasoline and motor oil, bromobenzene can also be formed when a water supply is chlorinated. It is a skin irritant and a depressant of the central nervous system in humans.

Bromobenzene is now unregulated but is a candidate for future regulation. The EPA called for monitoring to commence in community water systems by June, 1988.

CARBON TETRACHLORIDE

This volatile organic chemical is primarily used as a solvent. It is also used as a dry-cleaning fluid, a fire retardant, and in the production of pesticides. In animals, carbon tetrachloride is a proven carcinogen. It also causes damage to the developing embryo/fetus, the nervous system, and the liver and kidneys.

The SDWA requires regulation of concentrations of carbon tetrachloride in community water systems. The EPA called for monitoring to commence by June, 1987 (MCL = 5 ppb).

CHLOROTOLUENE

Chlorotoluene is used as a solvent and in the manufacture of dyes, drugs, and chemicals. It has been shown to be a mutagen and carcinogen in animal tests.

Chlorotoluene is now unregulated, but is a candidate for future regulation. The EPA called for monitoring to commence by June, 1988, for the two varieties found in drinking water: o-chlorotoluene and p–chlorotoluene.

DICHLOROBENZENE

Dichlorobenzene is used as a fumigant (in garbage cans and bathrooms); in mothballs; in the production of chemicals and dyes; and as a degreaser. About sixty million pounds were produced in 1975.

Dichlorobenzene is a suspected cause of hemolytic anemia, and is a carcinogen for animals. Tests on animals have also shown it to cause damage to the nervous system, kidneys, and liver.

Dichlorobenzene is now unregulated, but is a candidate for future regulation. The EPA called for monitoring to commence by January, 1988 (MCL = 75 ppb for para-dichlorobenzene).

DICHLORODIFLUOROMETHANE

Dichlorodifluoromethane is used as a refrigerant, aerosol propellant, and foaming agent. It causes heart disorders and cancer in animal tests. The EPA estimates a chronic dose of 1.9 ppb will cause one cancer death per 100,000 population.

Dichlorodifluoromethane was removed from the EPA's priority pollutant list in 1981. Monitoring requirements are set by state water quality agencies. It is difficult to detect.

DICHLOROETHANE (DCE)

DCE is used as a solvent and degreaser, in the production of paints and chemicals (such as vinyl chloride), and in the processing of ores. In animal tests, DCE has been found to be a mutagen and carcinogen, and to cause damage to the nervous system and developing embryo/fetus.

DCE is now unregulated, but is a candidate for future regulation. The EPA called for monitoring to commence by June, 1988 (MCL = 5 ppb).

ETHYLBENZENE

Ethylbenzene is used as a solvent, a component of gasoline (up to 20 percent of the total product), and in insecticides. It is also found in auto emissions.

Ethylbenzene causes kidney and liver disorders in test animals. It hasn't yet been adequately tested as a carcinogen in animals.

Ethylbenzene is now unregulated, but the EPA has proposed adding the chemical to the list of regulated substances that must be monitored in community water systems.

FLUOROTRICHLOROMETHANE (ALSO CALLED TRICHLORFLUORMETHANE)

Fluorotrichloromethane is used as a refrigerant (sold under the trade name Freon), in fire extinguishers, and as a cleaning agent. More than 300 million pounds are produced annually in this country. It has been shown to cause disorders of the heart in animal tests.

Fluorotrichloromethane is now unregulated, but is a candidate for future regulation. The EPA called for monitoring at the discretion of state water quality agencies by June, 1988. It is difficult to detect in water.

METHYL-TERTIARYBUTYLETHER (MTBE)

MTBE, a widely used component of gasoline, was added to the Drinking Water Priority List in 1988 after it was found in drinking water in many parts of the country. MTBE is a suspected carcinogen.

NITROBENZENE

Most of the nitrobenzene produced in the country is used to produce aniline, a chemical used in the production of dyes, rubber, and solvents. Nitrobenzene is also used in the production of explosives, and as a solvent. It is readily absorbed by the skin and by the lungs.

Nitrobenzene can be converted while in the stomach to compounds that can cause methemoglobinemia. Nitrobenzene's acute effects are on the central nervous system. Symptoms include fatigue, headache, vertigo, vomiting, severe depression and lack of energy.

Nitrobenzene is not regulated in drinking water.

POLYNUCLEAR AROMATIC HYDROCARBONS (PAHS)

PAHs are a family of toxic chemicals produced by the incomplete combustion of fuels in the presence of too little oxygen. They get into drinking water as the result of air-pollution-contaminated rain, from polluted runoff, and from industrial waste water. Benzo-a-pyrene is the most common PAH.

Animal tests have shown PAHs to be potent mutagens and carcinogens. There is strong evidence PAHs cause cancer in humans.

The SDWA requires regulation of concentrations of PAHs in community water systems. The EPA called for monitoring to commence by June, 1989.

POLYCHLORINATED BIPHENOLS (PCBS)

This family of more than 200 synthetic organic chemicals is used in the electronics industry and in electric power transformers. Water polluted with PCBs has been found in all parts of the country, with perhaps the worst contamination occurring in the lower Hudson River in New York as the result of years of dumping by General Electric.

PCBs cause damage to chromosomes, the nervous system, and the liver, and they have been found to be carcinogenic, teratogenic, and mutagenic in animal tests. PCBs are very persistent in human and animal tissues, and in the environment. PCBs break down into dibenzofurans, which are more potent.

The SDWA requires regulation of concentrations of PCBs in community water systems. The EPA called for monitoring to commence by June, 1988 (MCLG = 0).

STYRENE

Styrene is used in a wide variety of resins and plastics, including resins used to treat drinking water. Nearly one-and-a-half million tons are produced annually.

Styrene is toxic to the human nervous system and is mutagenic. Animal tests have shown chronic exposure to styrene can cause biochemical changes in the brains of test animals and damage their livers. Styrene is a carcinogen for some species.

Styrene is now unregulated, but is a candidate for future regulation. The EPA called for monitoring to commence by June, 1988, and has called for the addition of styrene to the list of toxics that must be monitored by community drinking water systems.

TETRACHLOROETHYLENE OR PERCHLOROETHYLENE (PCE)

Tetrachloroethylene is used as a solvent and a heat-transfer medium, and in the production of fluorocarbons. Water drawn from eight of ten water systems sampled in 1975 by the EPA contained traces of the chemical. More than 350,000 tons are produced annually in the United States.

PCE causes liver and kidney damage and central nervous system depression, and is a suspected carcinogen.

The EPA has recommended restrictions on the use of PCE because of its widespread occurrence in water and its apparent toxicity. The SDWA requires regulation of concentrations of PCE in community water systems. The EPA called for monitoring to commence June, 1987.

TOLUENE

Toluene is a product of refined petroleum or distilled coal tar. It is used in the production of chemicals, perfumes, dyes, solvents, drugs, TNT, and detergents, and is a component of gasoline (and of automobile exhaust). Nearly four million tons are produced annually in the United States.

Acute exposure to toluene causes a narcotic effect in humans. Chronic exposure in dogs results in nervous system intoxication, uncoordinated reflexes, and paralysis of the hind legs. Nervous system disorders have been reported in workers routinely exposed to toluene. Animal tests have resulted in an increased rate of miscarriage and cancer in some species.

Toluene is now unregulated but is a candidate for future regulation. The EPA called for monitoring to commence by June, 1988.

TRICHLOROETHANE (TCA)

Trichloroethane is used as a solvent, degreaser, and cleaner in a variety of industries. TCA is also used in the manufacture of computer chips. Leaking storage tanks in California's Silicon Valley have polluted with TCA an aquifer that provides drinking water to 20,000 people. Forty million dollars have been spent so far in an attempt to stop the spread of the pollution.

Acute exposure to TCA results in depression of the central nervous system. Chronic exposure produced liver and kidney damage in animal tests along with some signs of mutagenicity.

TRICHLOROETHYLENE (TCE)

This industrial solvent (and septic tank cleaner) has been found in drinking water in all parts of the nation. Tucson, Arizona; Denver, Colorado; and Woburn, Massachusetts, are among the towns where groundwater has been contaminated. Aquifers accumulate much higher concentrations of TCE because it evaporates readily from streams (and showers).

Animal tests have shown TCE to be a mutagen and a carcinogen in most species, and to cause liver and kidney problems. It has proven to be a nervous system depressant in both humans and animals.

The SDWA requires regulation of concentrations of TCE in community water systems. The EPA called for monitoring to commence June, 1987 (MCL = 5 ppb).

VINYL CHLORIDE

Vinyl chloride is used primarily in the production of polyvinyl-chloride (PVC) resins. It was also used as a propellant in aerosol cans until banned in 1974. Deterioration of plastic pipe is its primary source in drinking water. Water that has been standing in plastic pipe overnight or longer is most likely to contain vinyl chloride (and other impurities).

Vinyl chloride is a known human and animal carcinogen. It has also been demonstrated to cause disorders of the nervous system and brain in humans. Animal tests have shown the chemical to be a mutagen and to cause damage to the liver, nervous system, and lungs.

The SDWA requires regulation of concentrations of vinyl chloride in community water systems. The EPA called for monitoring to commence June, 1987 (MCL = 2 ppb).

XYLENE

Xylene is formed through the distillation of petroleum, coal tar, and coal gas. It is used in solvents, aviation gasoline, rubber cement, and the production of chemicals.

Xylene has been found to cause birth defects and damage to the nervous system, liver, and kidneys in animal tests.

The SDWA requires regulation of concentrations of xylene in community water systems. The EPA called for monitoring to commence by June, 1987.

PESTICIDES FOUND IN DRINKING WATER

ALACHLOR

Alachlor is the most widely used herbicide in the United States. More than eighty million pounds are used annually to kill weeds in corn, soybeans, and other crops. Alachlor is produced by the Monsanto Company under the tradename Lasso. An EPA survey found it in drinking water in Minnesota, Ohio, Nebraska, Maryland, and Pennsylvania, with the highest concentrations in Iowa. It is found in both ground and surface water. Alachlor has been proven to be a potent carcinogen in animals.

The SDWA requires regulation of concentrations of alachlor in community water systems. The EPA called for monitoring to commence by January, 1988 (MCLG = 0). Alachlor was reclassified during 1987 by EPA as a restricted-use pesticide. It was banned in Massachusetts starting in 1988, and its use was restricted in Canada after the herbicide was found in drinking water supplies.

ALDICARB

One of the most toxic, this carbamate pesticide is made by Union Carbide under the tradename Temik. It is used on potatoes, peanuts, sugar beets, citrus crops, and (especially) cotton. First found in groundwater in Suffolk County, New York, in 1979, aldicarb has since been found in aquifers in all parts of the country. On Long Island alone, more than 4,000 wells are contaminated.

Aldicarb is highly toxic to both animal and human nervous systems. It caused changes in the immune system of twenty-three women tested at a U.S. Center for Disease Control study.

The SDWA requires regulation of concentrations of aldicarb in community water systems. The EPA called for monitoring to commence by June, 1988 (MCLG = 9 ppm). Aldicarb has been banned in twelve states, but isn't regulated nationally. Two related compounds are being proposed by the EPA for addition to the list of toxics monitored in community water systems: aldicarb sulfoxide (MCLG = 9 ppb) and aldicarb sulfone (MCLG = 9 ppb).

ALDRIN AND DIELDRIN

These two chlorinated hydrocarbon insecticides were first criticized in 1962 by Rachel Carson in her book *Silent Spring* (as were their chemical siblings, chlordane, heptachlor, and endrin). She pointed out that dieldrin is five times as toxic as DDT when swallowed and forty times as potent when applied to the skin. The chemical's effects as a carcinogen, a teratogen, and a central nervous system toxin were becoming clear even at that time. By 1971 dieldrin was found in the tissue of 99 percent of Americans.

The use of aldrin and dieldrin were finally restricted by the EPA in 1974, but only after five years of litigation by the Environmental Defense Fund. Aldrin breaks down to become dieldrin once it enters the environment. Dieldrin is very persistent and takes up to fifteen years to break down. Dieldrin causes cancer in animals at the lowest doses tested, 100 ppb.

Concentrations of aldrin and dieldrin in drinking water aren't regulated.

CHLORDEONE (KEPONE)

A chlorinated hydrocarbon insecticide used on bananas, tobacco, and to control ants and roaches, chlordeone is very persistent once it enters the environment. It was sold under the tradename Kepone and manufactured by Life Sciences, Inc., of Hopewell, Virginia. Shellfish in the James River suffered an estimated $500 million in damage when Kepone residues dumped by the manufacturer found their way into the river. Groundwater in tobacco-producing areas is the most likely place to find Kepone contamination today.

Sales and production of Kepone were banned in 1975 when many of the employees of the company that produced it became seriously ill. Symptoms included nerve damage, memory loss, slurred speech, severe weight loss, sterility, and liver damage. It was found to cause liver cancer in animals. Concentrations of Kepone in drinking water are not regulated.

CHLORDANE

A chlorinated hydrocarbon insecticide manufactured by Velsicol Corporation, chlordane was the country's leading household insecticide in the 1960s and early 1970s. Chlordane and the related insecticide, heptachlor, have since been used only for termite control in this country. The two chemicals are found in the tissues of 95 percent of Americans.

Tests have shown chlordane causes cancer in animals. It is also suspected of causing disorders of the immune and nervous systems.

The SDWA requires regulation of chlordane. The EPA announced monitoring is required for all systems by June, 1988. Chlordane was banned for use on food crops by the EPA in 1975. All uses are banned in Massachusetts, New York, and Japan. Monitoring by the manufacturer in 1985–1986 revealed that chlordane and heptachlor infiltrated homes being treated for termites, resulting in a complete ban in 1987.

DICHLOROPHENOXYACETIC ACID (2,4-D)

A phenoxy herbicide, 2,4-D is the most widely used herbicide in the United States. Eighty million pounds are produced annually for use in 1,500 products, used primarily to kill weeds in lawns, gardens, and farms, and weeds and broad-leaf vegetation in forests. Concentrations of up to 70 ppb of 2,4-D have been found in Oregon streams as the result of aerial spraying of the herbicide on forestlands. It is also used to control plants growing in canals and bayous—even in water-supply systems! 2,4-D is especially lethal when combined with the herbicide 2,4,5-T (Silvex), as it was in Agent Orange, the herbicide used to kill trees and brush during the Vietnam War. It is found in water in all parts of the country.

The herbicide has been recognized since 1962 as playing a role in neuritis, paralysis, and chromosome damage. More recently it has been linked to neuropathy, changes in the central nervous system, miscarriages, damage to the developing fetus, and cancer in animals. An increased rate of miscarriages and birth defects among women living in the parts of the Oregon Coast Range sprayed with 2,4-D (and 2,4,5-T, in some cases) during the 1970s led to a temporary suspension of the spraying of herbicides on the country's national forests, a ban scheduled to end in 1988. Whole communities in the Coast Range were stricken with elevated rates of stomach and intestinal disorders while the spraying was still being carried out. Higher than average rates of cancer were also reported in the area.

A 1986 report by the National Cancer Institute says lymphatic cancers other than Hodgkins disease increased in farmers who use the chemical. Swedish researchers found soft-tissue sarcoma, Hodgkins disease, and non-Hodgkins lymphoma risk increased by phenoxy herbicide use. The SDWA requires regulations of concentrations of 2,4-D in community water systems. The EPA called for monitoring to commence by June, 1988 (MCL = 10 ppb).

ENDRIN

Endrin is another of the chemicals attacked by Rachel Carson in her 1962 book. She stated that endrin is fifteen times as potent as DDT to mammals and up to 300 times as potent to certain birds. Endrin is persistent in the environment, and can last more than a decade in most soils. Spraying to control cutworms in Montana wheat fields in 1981 led to contaminated water and waterfowl in much of the central part of the state. Although many restrictions have been put on the use of endrin, it can still be used on most crops.

In animal tests, endrin has been proven to be a carcinogen, to reduce fertility, and to impair the functioning of the central nervous system.

The SDWA requires regulation of concentrations of endrin in community water systems. The EPA called for monitoring to commence by June, 1989 (MCL = .2 ppb).

CHLOROBENZENE (ALSO CALLED MONOCHLOROBENZENE)

Chlorobenzene is used in the manufacture of pesticides and other chemicals, and in the production of dyes. Animal tests have shown it to be a mutagen.

Chlorobenzene is now unregulated but is a candidate for future regulation. The EPA called for monitoring to commence by January, 1988.

DDT (DICHLORODIPHENYLTRICHLOROETHANE)

A chlorinated hydrocarbon insecticide, DDT was first formulated in 1938. At the height of its use in this country in 1963, 176 million pounds were produced for use on 334 crops. DDT and the related chemicals DDE and DDD are today found in water and human and animal tissue in all parts of the country, even though it was banned in December, 1972. Its concentration in drinking water is not regulated.

DDT's persistence and the fact that it tends to accumulate in the fat of humans and animals are its primary health dangers. Although only moderately toxic, it can build up to toxic levels. DDT has been shown to be carcinogenic for mice, but not for rats.

DIAZONIN

This organophosphate pesticide is applied on a wide variety of commercial crops and in the home to control pests in the lawn, garden, and on pets. In 1971, 3.2 million pounds were used (about 2 percent of total insecticide use for that year). Animal tests have shown diazonin to be mutagenic and mildly teratogenic. Diazonin in drinking water isn't regulated.

DIBROMOCHLOROPROPANE (DBCP)

A nematocide found in aquifers underlying agricultural regions in many parts of the country (see the story in chapter 4), DBCP is especially prevalent in areas where citrus crops, grapes, or pineapples are grown.

DBCP has been shown to be a potent carcinogen and a mutagen

in animal tests. It also causes damage to the liver and kidneys. DBCP caused sterility in the majority of chemical plant workers handling it during the 1970s. Reproductive problems in females, including an increased rate of miscarriages, can also result from exposure to the chemical.

The EPA banned DBCP for all uses except on pineapples in Hawaii in 1979. SDWA requires regulation of DBCP concentrations in drinking water. The EPA called for monitoring in all systems to commence by June, 1988 (MCLG = 0).

DICHLOROPROPANE

Developed by Shell as a nematocide, dichloropropane is also used as a solvent, a degreaser, and an anti-knock additive in gasoline. Animal experiments have shown that the chemical causes depression of the central nervous system, that it is a mutagen, and that it causes liver and kidney disorders.

Dichloropropane is now unregulated, but is a candidate for future regulation. The EPA called for monitoring to commence by June, 1988 (MCLG = ppb).

DINOSEB

An insecticide and herbicide that is widely used on cotton and also on peanuts, soybeans, and potatoes, dinoseb is an active ingredient in 180 pesticides, none intended for home use. Dinoseb has caused birth defects and irreversible neurological and skeletal malformations in the offspring of exposed lab animals and sterility in male rats and mice.

The SDWA requires regulation of concentrations of dinoseb in community water systems. The EPA has called for monitoring to commence by June, 1989.

DIOXIN (2,3,7,8-TCDD OR TETRA DIOXIN)

Dioxins are a family of impurities produced in the manufacture of pesticides such as 2,4,5-T and pentachlorophenol. Tetra dioxin is the most toxic. It is one of the most toxic substances known. A single ounce is capable of killing hundreds of thousands of people. It was dioxin that was largely responsible for the cessation of the aerial spraying of Agent Orange in Vietnam (owing to concern about the birth defects it was apparently causing), and it was dioxin (as a contaminant in waste oil sprayed on gravel roads and on two horse arenas to settle dust) that prompted the permanent evacuation of Times Beach, Missouri, in 1983.

Dioxin in parts-per-*trillion* doses has caused cancer and birth defects in animals. It is both a mutagen and a teratogen, and is also toxic to the liver and nervous system. It causes skin disorders in humans.

The SDWA requires control of dioxin contamination in community drinking water systems. The EPA has called for monitoring to commence in June, 1989.

EPICHLOROHYDRIN (ECH)

ECH is used as a solvent for resins, gums, cellulose, and paints; in the manufacture of epoxy resins and synthetic glycerin; and as an insecticide. More than seven million pounds were produced in 1978.

In animals, ECH has been found to cause male sterility, to be a potent mutagen, and to be a carcinogen. It has also been shown to cause irreversible tissue damage when inhaled and to cause damage to the liver and kidneys.

The SDWA requires regulation of ECH's concentration in community water systems. The EPA called for monitoring to commence by June, 1988, for all systems.

ETHYLENE DIBROMIDE (EDB)

Developed by Dow Chemical in the 1920s as a gasoline additive, EDB was found to be an effective nematocide during the 1940s. EDB is chemically similar to DBCP. EDB has also been used as a preservative for stored grains and to control fruit flies. Its primary use today is as an additive to leaded gasoline (it helps to clean the lead out of the engine). EDB is more toxic to plants than the nematocide DBCP, and could not be used in fields containing crops.

The discovery of polluted wells in Florida's citrus belt during the early 1980s led to a suspension of its use as a nematocide in 1983. Use of EDB on food crops and grain was banned in 1984 after it was found to have polluted groundwater in agricultural areas in California, Arizona, Hawaii, and other states.

EDB was found to be a potent carcinogen in animals in 1975. It was also found that exposure to the chemical could cause sterility. As an animal carcinogen, EDB was found to be more potent than DBCP. The chemical has also been demonstrated to be a mutagen in animal tests.

EDB is now unregulated, but is a candidate for future regulation. The EPA called for monitoring to commence in areas prone to EDB contamination by June, 1988.

HEPTACHLOR

A chlorinated hydrocarbon insecticide manufactured by Velsicol Corporation, heptachlor is a derivative of chlordane and is closely related to the other cyclodiene pesticides: aldrin, dieldrin, and endrin.

The Hawaii Department of Health recalled dairy products in 1982 when it was discovered farmers on Oahu had been feeding their cattle pineapple leaves tainted with the pesticide, and that the feed had contaminated their cows' milk. In 1986 more than 100 Arkansas farms were quarantined when it was found that farmers were feeding cows heptachlor-tainted mash from a local gasahol plant, which had been making alcohol using grain treated with the chemical. Heptachlor was, again, detected in the cows' milk. Groundwater in Hawaii, much of the South, and the Southwest has been found to be tainted with heptachlor.

Animal tests have shown heptachlor to be a mutagen and a carcinogen, and to cause damage to the nervous system, the liver, and the developing embryo/fetus. The chemical breaks down into heptachlor epoxide, which is four times as toxic, once it enters the environment. A recent study revealed nine out of ten Americans' bodies contain residues of heptachlor epoxide.

The EPA has proposed adding heptachlor and heptachlor epoxide to the list of chemicals for which community water systems must monitor, although their concentration in drinking water is currently not regulated. Heptachlor was banned in 1978 except for the treatment of grain seeds and pineapples and for termite control. A complete ban was instituted for food crops in 1983, but the EPA is still allowing farmers and chemical wholesalers to use their existing stocks. Velsicol Corporation still manufactures the chemical for export.

HEXACHLOROBENZENE (HCB)

Used in the production of PCP and chlorinated hydrocarbon pesticides and as a fungicide, HCB is very persistent once it enters the environment and, like the chlorinated hydrocarbons, it tends to build up in human and animal tissue. Although its use as a fungicide is limited, it is usually found as an impurity in the pesticides that are made from it.

Animal tests have shown HCB to be a mutagen, a carcinogen, and a teratogen, and to cause nervous system damage.

The SDWA requires regulation of concentrations of HCB in community water systems. The EPA has called for monitoring for all systems to commence by June, 1989 (MCLG = 0).

HEXACHLOROCYCLOPENTADIENE (C-56)

The chlorinated hydrocarbon from which the cyclodiene pesticides (aldrin, dieldrin, endrin, chlordane, heptachlor, and heptachlor epoxide) are synthesized, C-56 is most frequently found in water near chemical dumps related to the production of cyclodienes. Typical are those in Niagara County, New York (where Hooker Chemical dumped an estimated *20,000 tons* of C-56), and the Velsicol Chemical Corporation dump in Toone, Tennessee.

Acute exposure to C-56 causes respiratory distress, nausea,

memory loss, eye and skin irritation, and liver and kidney abnormalities. Little testing has been done to determine its chronic effects.

The SDWA requires regulation of concentrations of C-56 in community water systems. The EPA called for monitoring to commence by June, 1989. Exposure to more than 1 ppb is expected to be dangerous to humans, but no standards have yet been set.

LEPTOPHOS

This organophosphate insecticide is one of the few that is as persistent in the environment (and in human and animal tissues) as are the chlorinated hydrocarbons. It was produced by Velsicol Chemical Company for export to fifty countries during the early 1970s under the tradename Phosvel. Though the EPA in 1974 granted tolerances for the product's use on lettuce and tomatoes, a preliminary step to allowing Phosvel's use in the United States, evidence of its exceptional toxicity forced the agency to cancel the tolerances and institute a ban on its production in 1976. The American public's only exposure to Phosvel was in residues on imported crops, primarily tomatoes from Mexico (where one-quarter of the tomatoes consumed in this country are grown).

Serious effects on the health of the chemical workers that produced Phosvel played a large part in the banning of the insecticide. Workers at Velsicol's Bayport, Texas, plant, one of eight where Phosvel was produced, began to be stricken with a variety of disorders including blurred vision, dizziness, amnesia, tremors, bladder dysfunction, and loss of sensation, symptoms that are similar to those of multiple sclerosis. Several were permanently disabled by a disorder known as "spastic paralysis of the lower extremities." It was later found that the symptoms were the result of delayed neurotoxicity, a permanent degenerative disorder of the nervous system caused by leptophos and a few other organophosphates. At the same time, in Egypt, more than 1,000 water buffalo that had been working in fields treated with Phosvel suffered paralysis of their hind quarters and several field workers reportedly died.

LINDANE

Lindane is one of the country's most widely used household insecticides, although the EPA has been reviewing its registered uses since 1980. It is used in aerosol insecticides; flea collars and soaps for pets; in moth balls; and as a component of floor wax. Lindane is very persistent once it has entered the environment.

Lindane has been demonstrated in animal tests to be a potent carcinogen, to cause miscarriages as well as birth defects, and to damage the central nervous system. Aplastic anemia can follow acute exposure to Lindane in humans.

The SDWA requires regulation of concentrations of Lindane in community water systems. The EPA called for monitoring to commence by June, 1988 (MCL = 4 ppb).

METHOXYCHLOR

A close relative of DDT, methoxychlor has seen increased use since DDT was banned. It is much less persistent than DDT. An estimated ten million pounds were produced in 1975. It is registered for use on eighty-seven crops, including most fruits, vegetables, and grains; beef and dairy cattle; and pigs, goats, and sheep. It is also found in home and garden insecticides.

Methoxychlor's toxicity to humans and animals is relatively low. However, chronic exposure to the chemical can cause liver and kidney damage.

The SDWA requires regulation of concentrations of methoxychlor in community water systems. The EPA called for monitoring to commence by June 1988 (MCL = 100 ppb).

NAPHTHALENE

Naphthalene is used in mothballs (therefore it is present at landfills) and in the production of dyes. Anemia and disorders of the liver and kidneys can result from chronic exposure to it.

Naphthalene is now unregulated, but is a candidate for future regulation. The EPA has called for monitoring at the discretion of state water quality agencies.

PARATHION

One of the most widely used and most toxic organophosphate insecticides, methyl parathion has been shown in animal tests to be a mutagen and a teratogen. Some research has shown it can cause long-term brain damage in children. In the human body, it is converted to paraoxon, which can cause nerve damage. Like most organophosphates, it can pass through the skin. Parathion concentrations in drinking water are not regulated.

PENTACHLOROPHENOL (PCP)

PCP has been used as a wood preservative, a herbicide, and an insecticide since 1936. About 25,000 tons are produced annually. PCP is best known under one of its tradenames, Penta. Post and pole plants and railroad tie plants are a leading source of groundwater contamination with PCP.

PCP has been demonstrated to be a teratogen in animals. The fact that the traces of dioxin that are produced when PCP is synthesized are usually present in the finished product is the primary reason for concern about toxicity. (Dioxin has been shown to be a mutagen, a teratogen, and a carcinogen in animal tests.)

The SDWA requires regulation of concentrations of PCP in community water systems. The EPA called for monitoring to commence by June, 1988.

TRIAZINES

The four triazine insecticides—atrazine, simazine, propazine, and cyanazine—are used primarily on corn, sorghum, and sugar cane fields before the crops sprout. Atrazine is the most widely used. Tests have so far shown all four to be of relatively low toxicity to humans and animals.

The SDWA requires regulation of concentrations of simazine and atrazine in community water systems. The EPA has called for monitoring to commence by June, 1989.

TOXAPHENE

The insecticide toxaphene is used more and understood less than any of the chlorinated hydrocarbons. The reason is its complex structure. It is composed of more than 175 chemicals; the chemical structure of only ten is known. Tests have shown animals' kidneys and liver can degenerate as the result of chronic exposure to toxaphene.

The SDWA requires regulation of concentrations of toxaphene in community water systems. The EPA called for monitoring to commence by June, 1988 (MCL = 5 ppb).

2,4,5-T (TRICHLOROPHENOXYACETIC ACID)

2,4,5-T is a chloriphenoxy herbicide like its chemical cousin, 2,4-D, the other component of Agent Orange. Spontaneous miscarriages by women who lived in the Oregon Coast Range forests, where 2,4,5-T was sprayed during the 1970s to kill broadleaf vegetation, brought public pressure for restrictions, which were instituted in 1979. Fruits, rice, and range land were exempt from the ban.

The dioxin contaminants present in the product are the main reason for 2,4,5-T's exceptional toxicity. Experiments in the late 1960s revealed that the herbicide caused miscarriages and birth defects in test animals. Neurological damage, cancer, and damage to the kidneys and liver have also been documented in test animals exposed to 2,4,5-T. 2,4,5-T is now unregulated but is a candidate for future regulation.

RADIOACTIVE MATERIALS

The atoms in any radioactive material are constantly disintegrating and emitting particles and electromagnetic waves that can cause damage to whatever they hit. Alpha and beta radiation are both charged particles. Gamma rays are pure energy, since they have no mass. An x-ray is an artificially produced gamma ray.

Radiation is harmful because it can "ionize" living matter, making permanent changes in its atomic structure. Cancer, damage to developing cells in the fetus and growing children, and mutations of chromosomes are among the chronic effects of exposure to ionizing radiation. Concentrations of radioactive materials in drinking water aren't high enough to cause acute effects.

The three kinds of radiation vary in their ability to affect the body. Alpha radiation is highly ionizing but can't penetrate tissue. Beta radiation is more capable of penetration but somewhat less ionizing. Gamma radiation is the least ionizing but the most penetrating. Alpha and beta rays are normally a health threat only if swallowed, inhaled, or absorbed through the skin.

Radiation can have a dramatic effect on human health. A 1987 study at South Dakota State University found 29 percent of cancer in people under twenty in that state is caused by natural radioactivity and that 89 percent of cancer deaths among South Dakota residents under thirty-five were not related to lifestyle.

Radiation in water is measured in *picocuries per liter* (pCi/L). One *curie* is the amount of radiation emitted by one gram of radium. A *picocurie* is one-trillionth of a curie. A *roentgen* (pronounced "renkin") is the amount of x-ray (or gamma-wave) energy required to produce two billion ions in one cubic centimeter of air. A *rem* (roentgen equivalent in man) is a unit designed to measure the effects of alpha and beta waves and other radioactive sources that don't behave like gamma waves. A *millirem* (mR) is one-thousandth of a rem.

There are twenty states where the total radiation present in groundwater is up to 1,000 pCi/L. The EPA estimates that an individual who consumed water with this level of contamination for seventy years would run a 1-in-10,000 risk of contracting cancer. For a 10,000 pCi/L concentration, the agency estimates the chance of getting cancer rises to 1 in 1,000.

ALPHA PARTICLES

Alpha particles are emitted by the majority of radioactive materials, including radium-226 and the decay products of radium-228. Alpha radiation tends to accumulate in bones and can cause bone cancer.

The SDWA requires regulation of alpha particles in community water systems. The EPA has called for monitoring to commence by June, 1989 (MCL = 15 pCi/L).

BETA PARTICLE AND PHOTON RADIOACTIVITY

Beta particles are emitted by radium-228 and a few other radioactive materials. The materials found in surface water that emit beta radiation are primarily the result of atmospheric nuclear weapons testing. Groundwater contamination has occurred in areas where underground nuclear tests have been conducted. Several private wells in western Colorado were poisoned as the result of experimental detonations designed to stimulate natural gas production during the 1970s.

The SDWA requires regulation of concentrations of beta particles in community water systems. The EPA called for monitoring to commence by June, 1989 (MCL = 4 mR per year).

RADIUM

Radium, a decay product of uranium, is found primarily in groundwater, with the highest concentrations in Iowa, Illinois, Wisconsin, and Missouri. Radium-226 and radium-228 are the most commonly encountered forms in drinking water.

The SDWA requires regulation of concentrations of radium in community water systems. The EPA called for monitoring to commence by June, 1989 (MCL = pCi/L).

RADON

Radon is a radioactive gas produced by the breakdown of radium. A recent study found that in the Northeast, 50 percent of indoor air pollution by radon comes from water. Washing machines and showers are the primary source. There is an unusually high incidence of radon in the area, especially in the New England states. Radon concentrations in surface water are generally about 1 pCi/L. Concentrations in groundwater can be thousands of times higher. Mineral or spa waters may contain up to 500,000 pCi/L.

The SDWA requires regulation of concentrations of radon in community water systems. The EPA called for monitoring to commence by June, 1989 (MCL = 5 pCi/L).

URANIUM

Uranium gets into water from rocks in aquifers and from the tailings of uranium mines and mills. Two hundred million tons of uranium tailings are located around twenty-six mills in New Mexico, Wyoming, Utah, Colorado, Washington, and Texas. Uranium is also found in gypsum tailings. Radium-226 is also present in the tailings. Decaying uranium is a source of alpha particles.

The SDWA requires regulation of concentrations of uranium in community water systems. The EPA has called for monitoring to commence by June, 1989.

As you can see even from this *partial* list, the number of contaminants found in drinking water—and their toxicity—is staggering. The only way of being sure some (or all!) of the toxic substances reviewed aren't in *your* drinking water is through action: *yours*. Chapter 8 will outline steps you can take to *ensure* the quality of your water. The following chapter will tell you how to recognize a water source that is likely to be of good quality, and where the best average water quality in the nation is to be found.

7 · Where Is the Good Water?

As we've seen, bad water is really everywhere. Is there no good water left? And if there is, where is it, and how can it be recognized?

Luckily, although supplies of good water are dwindling, there *are* many areas where water quality is comparatively good. Traces of pollution can be found virtually everywhere, but in some locales, at least, it isn't necessary to take out a life insurance policy before drinking the local liquid! And even in places where most drinking water carries dangerously high concentrations of pollutants, there are sometimes individual water-supply systems of good quality.

In general, the best drinking water is found near the *source* of streams, and in lakes and aquifers that are well removed from heavily developed areas. This chapter will tell you what parts of the country have the best water and what some of the characteristics of a good water source are. Chapter 8 will give you more specifics on how to evaluate the water in *your* home for potential pollution by evaluating the watershed from which it comes.

GOOD SURFACE WATER

Water that flows out of sparsely populated hills or mountains where relatively little mining and logging have occurred is the water most

likely to be safe to drink. Such a source has simply had few
opportunities to become polluted.

The taste of water is the property we tend to pay the most
attention to; taste is an important component of what makes water
"good." It's hard to beat the flavor of water that has recently been
cascading down a mountain stream. It is rich in oxygen, which
improves its taste, and contains fewer minerals than most groundwater. Volatile toxics vaporize from the tumbling water, which is
further cleansed when it is forced through the streambed's sand and
gravel. Highly mineralized groundwater, while not necessarily bad
for health, has a "hard" taste.

Rattlesnake Creek, which was first diverted to provide Missoula,
Montana, with drinking water more than a century ago, is a good
example of such a stream. The creek drains the rugged south side
of the Rattlesnake Mountains, just north of town. It originates in a
designated wilderness area and passes through a national recreation area on the short journey to its confluence with the Clark Fork
River in Missoula. The Rattlesnake's watershed is managed to
maximize opportunities for recreation, to support wildlife, and to
protect water quality. Other than a corridor along the lower creek
where motorcycles and snowmobiles are allowed, access to the area
is only by foot and bicycle in the summer, and by skis and
snowshoes in the winter. Although there has been some logging in
the area above the reservoir where the city's water supply is
withdrawn, the Rattlesnake's headwaters have no history of widespread timber harvesting or mining.

Drinking water from Rattlesnake Creek has been of excellent
quality, good enough to be used with only light chlorination and no
filtration for decades. The first known health problem resulting
from its use was in 1983, when an outbreak in Missoula of the flu-
like disease giardiasis forced the local water utility to close down
the Rattlesnake water intake and to make up the resulting deficit
with groundwater. The aquifer underlying the Missoula Valley had
supplied about half the city's water before the closure.

Giardiasis is caused by an intestinal parasite, giardia lamblia.
As mentioned in the previous chapter, it has been referred to as the
"backpackers' lament" because the cysts that cause the disease are
most likely to be found in cold mountain streams across the United

States. A giardia parasite, like many other parasites and disease organisms, encases itself in a membrane sac when it is ready to go into its cyst phase. The coating protects the dormant organism until conditions are favorable for its re-emergence.

The giardia cysts in Missoula's drinking water were traced to a beaver that had set up housekeeping in the creek just upstream from the reservoir where the city's water is withdrawn from Rattlesnake Creek. Although the giardia parasite can live in the intestines of a variety of mammals, including dogs, horses, and many wild animals, in lakes and streams water-dwellers such as beavers and muskrats are its primary source. One beaver can release *one billion* giardia cysts *per day* into the water in which it lives.

When a giardia cyst is ingested in drinking water, it sheds its protective sheath and attaches itself to the intestines, where it absorbs nutrients and multiplies. When enough of the parasites become established on the intestine's walls, giardiasis develops. Diarrhea and weakness are the predominant symptoms; it is like a flu that won't go away. Fortunately, many people are immune.

The cyst can survive longest in cold water like that found in Rattlesnake Creek and similar mountain streams across the country, many of which are used for drinking water. The water drawn from such streams is some of the best available, orders of magnitude better than that of rivers like the Ohio, where hundreds of contaminants are normally found, but health-threatening pollution such as the giardia infestation of the Rattlesnake still occur.

Water from Rattlesnake Creek won't be used again until an activated carbon filtration system is built, at an estimated cost of $3 million, or until the city's water utility gets an EPA exemption to the filtration requirement by removing the beavers from the creek. Removal of the beavers, however, is opposed by many residents of the area. Community water systems were required to start monitoring for giardia during 1988.

Clearly, some of our finest "good water" is vulnerable. Don't assume that because your water comes from a high-mountain stream that it is absolutely pure. Even it may need some careful attention.

GOOD GROUNDWATER

It's a bit more difficult to characterize an ideal source of ground-water. The aquifer that became Missoula's sole source of drinking water with the closure of the Rattlesnake Creek water supply, while not pristine, can serve as a model. Its flow is a virtual torrent by groundwater standards—water moves through the aquifer at speeds of up to several feet per day. This rapid flow rate and the predominantly good quality of the water that recharges the aquifer help to remove any contaminants that may find their way into the ground-water. Although there have been numerous instances of aquifer pollution in the valley, they have been minor in comparison to the widespread contamination prevalent in many parts of the country.

The Missoula Valley aquifer has provided all the town's drinking water since the giardia infestation in 1983. Much of the water that flows into the aquifer comes from the unpopulated mountains that surround the town and is, as a result, of good quality. The water pumped from the aquifer is pure enough to be used without treatment. The valley's relatively light population and the small number of farms and industries located there protect the aquifer from serious contamination.

But again, even in a place like Missoula, there are problems with groundwater quality. The Missoula Valley is underlaid by about 2,500 feet of sand and gravel deposited over the ages by streams. Layers of water-bearing stream sediments are separated by thin layers of clay, which are largely impervious to the passage of water. Although groundwater can be found even at the deepest levels, it is only in the uppermost layer, within 200 feet of the surface, that groundwater is of good enough quality and adequate flow to be used for drinking water.

The top layer of water-bearing sand and gravel is "unconfined," meaning that there is no impervious layer between its top and the ground's surface. This means toxic materials from the surface can easily filter down into the groundwater. The large grain size of the sand in the area makes the penetration of toxics even easier. As a result, leaking underground storage tanks, waste dumps, accidental spills, and pesticide use are especially threatening.

Septic tanks, which can pollute groundwater even under the best of circumstances, constitute an even greater threat to an aquifer like Missoula's. A 1986 study by the University of Montana showed that most of the outflow from a new, properly installed septic tank soaks directly down into the ground instead of being carried out into the drain field to evaporate as it is supposed to. The Missoula Valley's sands are simply too coarse to keep the water near the surface long enough for it to evaporate. As a result, elevated nitrate levels are detected in those parts of the valley where most homes use septic tanks. There is also an increased threat of contamination from the many solvents and other toxic compounds that are commonly used around the home, then flushed down the drain.

Another potential source of groundwater contamination in Missoula is the Clark Fork River, which flows through the valley. The river has been severely contaminated with heavy metals released during a century of extensive mining and smelting at its headwaters in the Butte, Montana, area.

Approximately seven million cubic yards of sediments, containing tons of arsenic, cadmium, lead, and zinc, have accumulated in the Milltown Reservoir on the river a few miles upstream from Missoula. An estimated 1,700 *tons* of arsenic and 27,000 tons of zinc are contained in the sediments. The reservoir was named a priority site in the Superfund toxic waste cleanup program in 1981 when arsenic was discovered to have contaminated community water wells in nearby Milltown. More water from the Clark Fork can be expected to enter Missoula's aquifer now that it supplies all the town's water, since groundwater levels have fallen as a result of the increased pumping. Little is known, however, about what effect this has had on groundwater quality.

As can be seen from the example of Missoula, even in a town that has some of the purest water in the nation, there's no such thing as a foolproof source of clean water. But there are many areas, like Missoula, where average water quality is good compared to that in other parts of the country. Read on to find out where some of those areas are.

WHERE IS THE GOOD WATER?

THE NORTHEAST

Water in mountainous parts of the Northeast—the Catskills and the Adirondacks in New York, the Green Mountains in Vermont, the White Mountains in New Hampshire, the Longfellow Mountains in Maine, and several smaller ranges—is of generally high quality.

Except for the 600,000 Queens residents who buy groundwater (of poor quality) from a private water utility, New York City residents' drinking water comes from upstate reservoirs, which generally supply water of good—but unfortunately, diminishing—quality. The city's supply of water is adequate, in non-drought years, with an annual average of forty inches of rain falling on the watersheds that furnish the *1.5 billion gallons* of water used *daily* in New York City. The city's water system serves a total of ten million people.

The headwaters of the Delaware provide about half the city's water, 40 percent comes from the Catskills, and another 10 percent from the Croton System, in Putnam and Westchester Counties fifty miles north of town. The water that flows into New York through the system's two supply tunnels comes from drainages covering a total of 2,400 square miles. The existing supply lines are now barely able to keep up with the city's demand for water, even though both carry up to 60 percent more water than they were designed to transport.

A gargantuan third pipeline that will be capable of carrying *all* the city's water was started in 1968. Construction is still in progress, with completion expected around the end of the century. The estimated price tag of $5 billion represents the largest project ever totally financed by the city. A *twenty-four-foot diameter* pipe is being used. Completion of the first fourteen miles of the new pipeline, slated for 1990, will allow the shutdown of the lower part of the city's existing water-supply tunnels for the first time since their respective completion in 1917 and 1937, allowing officials to see what effect more than half a century of use has had on them.

Although much of the surface water drawn from near the headwaters of the northeast's streams is of basically good quality, it is increasingly vulnerable to pollution from a multitude of sources. Development in the drainages from which New York City's water originates is an ever-growing threat to water quality. More people, septic tanks, businesses, and all the other by-products of growth mean deteriorating quality of streams, lakes, and aquifers. The ever-greater use of salt to de-ice roads in winter is causing a rise in the salinity of New York's City's water, and acid precipitation is further threatening the water's quality.

Deteriorating water quality resulting from local growth is most severe in the Croton watershed. An $11.5-million demonstration filtration plant was completed in 1987 to clean part of the Croton water, and a $320-million plant in the Bronx to filter the rest is slated for the 1990s.

Outbreaks of giardiasis have been reported in other northeastern cities with similar water systems, but so far no such problems have been encountered in New York City. The city is expected to ask the federal government to exempt it from a requirement to install filtration equipment capable of removing the parasites.

But even with these threats, New York City's drinking water (aside from the borough of Queens) remains consistently some of the best supplied to any major American city. Several communities in the northeast that draw drinking water from lightly developed mountains have similarly good water. Long Island communities, as previously noted, do not fare nearly as well.

Although some of the worst groundwater contamination in the country is in urban portions of the Northeast, aquifers in the more rural parts of the region are in relatively good shape. The best groundwater is found in most of the same areas where the surface water is in good condition—in or near lightly populated mountains and hills. Leaking underground storage tanks, garbage dumps, and septic tanks are leading threats to the quality of groundwater in such areas. Agricultural chemicals and nitrates from fertilizers pollute aquifers underlying farmlands.

THE ROCKY MOUNTAIN STATES

Lightly populated areas in the Rockies also have, on average, some of the nation's best water. Areas where there has been extensive

mining or where forestlands have been heavily treated with herbicides are exceptions. Logging and extensive cattle grazing also
degrade the quality of many mountain streams, primarily by increasing erosion. Giardia cysts are found widely in Rocky Mountain streams.

In Denver, where, as we've seen, there are many sources of water
pollution, high-quality surface water captured along the Front
Range of the Rockies and on the Western Slope (from the headwaters of the Colorado River) is blended with other sources to
improve average water quality. Although the heavy mining that has
occurred in Colorado's mountains has ruined the water in some
areas, there is still plenty of high-quality mountain water in the
state.

Idaho and Montana have some of the best water in the Rocky
Mountain region because of their light population and limited
development. The condition of drinking water in Missoula, Montana, described at the beginning of the chapter, is fairly typical—
there is some contamination, but it is generally far less than is
typical in more heavily populated areas. Mining and ore milling
cause the most serious contamination of the mountain states'
streams. Mining centers such as Butte, Montana, and Kellogg,
Idaho, have made a remarkable recovery from the toxic abuse they
have suffered in the past, but the scars of that abuse live on in the
form of seriously polluted ground and surface water.

Some damage is widespread, as is the pollution of the Clark Fork
River from its headwaters in Butte to its mouth in northern Idaho's
Lake Pend Oreille, but most streams are of good quality near their
sources (if they don't have the bad luck to emerge in Butte—as the
entire city is a Superfund site!). The degradation of watersheds
resulting from the combined effects of logging and grazing has also
reduced water quality in many parts of the Rockies. Contamination
from herbicides used on woodlands is found in a few areas, but
isn't as threatening as it is in the forests nearer the Pacific.

The best average *groundwater* quality in the Rocky Mountain
states is also found in Idaho and Montana. Although agriculture is
one of the primary pursuits in both states, the yield from the often-
thin soils in the frequently severe climate makes it hard to justify

the scale of pesticide and fertilizer use practiced in other, more productive agricultural regions. Although nitrates can be found in some of the area's groundwater, they are more likely to come from septic tanks and animal feedlots than from heavy use of fertilizers, as in the Midwest.

Groundwater contamination from mining and milling isn't as extensive in Idaho and Montana as it is in the southern part of the Rocky Mountain region. Tailings from mines and smelters and the networks of tunnels that may cut through aquifers are both threats to groundwater quality.

THE PACIFIC NORTHWEST AND NORTHERN CALIFORNIA

The Cascade Mountains in western Oregon and Washington, the Sierras in northern California, and the coastal mountains from San Francisco to the Canadian border are all sources of relatively good water. As with the Rocky Mountain states, mining and logging have caused serious contamination in some, but certainly not all, areas.

When a coniferous forest is clear-cut in this wet coastal climate, fast-growing deciduous trees (especially alders) and undergrowth quickly take over. During a period of decades, if undisturbed, the leafy vegetation will mature and be slowly replaced by a new generation of coniferous trees such as fir, redwood, or cedar. Herbicides are used in many of the region's forests to kill the broad-leaf vegetation and encourage quicker growth to conifer seedlings, thereby cutting the delay until the next harvestable timber is available.

Traces of the herbicides used invariably end up in local water supplies, and are one of the principle pollutants found in what is predominantly a region of good water. Contamination has been most severe in the coastal mountains, where the growth of broad-leaf vegetation is most prolific, but streams flowing from the Sierras and the Cascades also are affected. Mining, although a problem in some areas, is less predominant in the coastal mountains than is the case in the Rockies. Increased development of mountain communities is increasing water pollution in some areas.

It is a sad fact that we are now at a point where bad water is the rule and good water the prized exception. Good water is an increasingly valuable—if dwindling—economic and health asset, drawing more and more people into those areas that have it, and ironically jeopardizing what's left in the process. The only solution is to clean up the water *everywhere* and, in the meantime, to take measures to safeguard our own personal water supplies as best we can.

8 · Cleaning Up *Your* Water

N ow that you have some idea of where the good water is and the bad water is, how do you assess the condition of the water in *your own home?* Even if you are fortunate enough to live in an area with little pollution, you may not be home free. As we've seen, even the best water can be contaminated en route to the kitchen sink. Whether the contamination comes from the city water distribution system, the drinking water treatment plant, or your own home's plumbing system, the result is the same: bad water.

Many have been assured by the salesman of a water-purification device that "this will take care of everything," but such claims are too often meaningless. Every method of purification has advantages and limitations—no single method can remove *everything* toxic from seriously polluted water. Two or more stages of cleaning usually must be combined to remove most of the toxic elements from most of the water; even then, some contaminants will remain.

Before you decide how best to clean up your water, however, you must know what's in it, and finding that out can be more difficult than it sounds. It can also be expensive. *Complete* analysis of the hundreds of contaminants that *can* be present in a single sample of drinking water can cost *thousands* of dollars. Fortunately, such an extensive analysis is seldom needed. You can order a test that identifies the sixty or so most common contaminants; it will cost about $200, and will suffice in most cases. Or, you may opt for simpler tests and still improve your water. The cost of testing varies, so it pays to shop around. See the sidebar for a list of labs that test water from all parts of the nation.

National Water Testing Laboratories

Brown and Caldwell Laboratories	415-428-2300
	818-795-7553
Montgomery Laboratories	818-796-9141
Wilson Laboratories	800-255-7912
(Kansas residents call:	800-432-7921)
Suburban Water Testing Laboratories	800-433-6595
(Pennsylvania residents call:	800-525-6464)
Water Test Corporation	800-426-8378
National Testing Laboratories	800-458-3330

You'll need to know at least what kinds of contaminants *may* be present in your water before you have it tested. Looking for the toxic components in your water can be like looking for a needle in a haystack. The more you know about just what you're looking for, the more successful the outcome, and the less extensive and expensive the testing.

This chapter will tell you how to determine which toxic elements are *most likely* to be in your drinking water by evaluating the water's source, treatment, and distribution. It will also tell you where you can get assistance and how to take a water sample, and it will describe the principal types of water-purification devices now available. Reviews of representative purification products will close the chapter.

FIRST FIND OUT WHAT *MAY* BE
IN YOUR WATER

Finding out where your drinking water comes from is the first step toward cleaning it up. Once you know that, you'll have some idea of the *range* of contaminants that may be in it.

If your water comes from a well in the back yard, your task is somewhat simplified, but it still may be hard to find out much about the aquifer your well's water comes from. Too often, studies of the quality and motion of groundwater come only after the discovery of seriously contaminated wells. Without such studies, it is impossible to predict accurately the movement of polluted water within an aquifer. Those using water from such an aquifer are left in the unsettling position of not knowing whether their drinking water is threatened until contamination is actually detected in a water sample (and there is usually a lapse between contamination and detection). With groundwater supplying half the water used by cities and virtually all rural drinking water, the dearth of information about the nation's aquifers is a serious health threat.

If you get your water from a private well, health experts recommend having the water tested at least once a year for bacteria. The test will cost about ten dollars.

If your water comes from a large community system, even determining its source can take a little digging. Such a system typically draws from an ever-changing mix of water sources. A variety of lakes and rivers may supply more water during wet years, with aquifers used more heavily during dry spells. New water sources may be tapped while others are temporarily or permanently closed because of contamination. Changes in the way drinking water is processed at the treatment plant and in the system of pipes and valves through which it is delivered will also affect quality.

Luckily, abundant help is available. Your public library is a good place to start. An index of a local newspaper will lead you to articles about the water system, its construction and expansion, and stories about known water-quality problems in your area. Before visiting the library, be sure to consult the bibliography in this book. See Resources Section A for a *list of hotlines* where you can get quick answers to questions regarding water quality.

The water company, the local health department, and the state's water quality and health agencies (Resources Section C) are the best places for getting specifics about your drinking water's sources and purity. The water utility will have maps of both the supply and distribution ends of the community's water system, the results of tests of each water source, and details of the water treatment process. The water department is the best place to find out about the general condition of the water system and about plans to improve it.

The Safe Drinking Water Act *requires* monitoring of "community" water systems—those that have fifteen or more service connections or that supply an average of twenty-five or more people with water for more than sixty days during the year. Seventy percent of the 206,000 public water systems in the United States do not qualify as a community water system because they serve too few people or because they are used only seasonally. It has been estimated that there are 850,000 water systems in the country with between two and fourteen connections, which are not considered community water systems.

If monitoring is required and *if* your local community water system is in compliance with those requirements, a record of the substances regulated under the Act will be available (see chapter 7 for details). Be aware, however, that a survey by the congressional Government Accounting Office (GAO) concluded that more than half of the nation's community water systems fail to meet the Act's testing requirements.

The health department in your community may be able to perform some testing, primarily for bacterial contamination, either free or at a reduced rate. You should also be able to find out about known and suspected sources of drinking water pollution in your community from local health department officials. If you suspect your water is being polluted, the city or county health department and the state water quality agency are usually the first places to go for help. A list of *certified water-testing labs in your state* should be available from the state or the health department. These labs may know exactly what to look for in your water since they will already have tested many samples from your city, community or rural area.

In your search for information about the quality of your water, don't overlook your neighbors (especially if you're new to the area). They can often give you insights into the history of local water quality and the sources of local water pollution. You also have a common interest with your neighbors in maintaining the quality of your drinking water. Sadly, a high incidence of disease in an entire neighborhood has too often been the way in which residents of a community have found out about bad water quality. Obviously, it's best to take action before that happens.

You probably won't have to contact *all* these sources to find out what you need to know, but before your search is over, there are several questions for which you need answers.

If groundwater supplies part or all of your drinking water:

- What is the nature of the aquifer from which it comes?
- In what direction does its water flow and where are its "recharge" areas; that is, where does water from the surface soak into the ground and enter the aquifer?
- Is the aquifer "unconfined," like Missoula's (discussed in the previous chapter), and therefore open at all points to pollution from above?
- Are there any injection wells in the area, and, if so, *what* do they inject and *where?*
- Have there been any problems with leakage?
- What are the known and suspected sources of pollution in your area?
- What kind of pollution do they produce? Specifically, are there any toxic waste disposal sites or municipal dumps in the areas where your water originates or overlying your aquifers?
- Are insecticides, herbicides, and fertilizers widely used in your area? Which ones?
- Have solvents gotten into groundwater from industry or from domestic septic tanks within the area?

The nearest university and local and national citizen's groups concerned with water quality may also be able to help. See Resources Section A for listings of associations, foundations, businesses, environmental groups, and government agencies that offer information about drinking water quality. If there is a university in your area that offers courses in hydrology and hydrogeology, studies of local aquifers may be available through the school's library.

The listings in Resources Section D of Superfund sites and in Resources Section B of the total toxic wastes produced in each state will give you an idea of the overall scale of potential contamination in your area. The Thomas Register at your library and your local Chamber of Commerce are also good sources of data on the number and kind of manufacturing businesses in your area.

It is usually a good idea to test for some of the principal components of gasoline if you use groundwater *anywhere* in the United States, since spills and leaking storage tanks are a universal threat. The human nose is as sensitive as the most sophisticated test equipment in sniffing out gasoline contamination, so you'll probably know if you have a problem. Some people have become so accustomed to the odor, however, that they have concluded that this smell is a natural component of their water. That is a *very* hazardous conclusion.

Don't be surprised if the "facts" you receive from the sources you consult don't agree. You'll want to consider *all* the possibilities at this point so the lab that tests your water won't miss anything important.

POLLUTANTS FROM THE TREATMENT AND DISTRIBUTION SYSTEM

Toxics picked up in treatment and distribution systems must also be considered. If chlorinated surface water is part of the mix used by your community's water system, the trihalomethanes we discussed earlier, such as chloroform, are likely to be present (see the toxics listed in chapter 6 for a discussion of some of the primary chemicals used as disinfectants in water treatment plants, and their effects).

Aging asbestos-cement water distribution lines, corroding water pipes, anti-corrosion treatments, and toxic substances that leak into cracked water pipes while the water is turned off are among the major elements that can pollute the fresh water being distributed in a water system, no matter how pure it may have been to begin with. Corrosion of deteriorating water-supply pipes can taint water supplies with metals such as zinc and lead. Plastic water mains can pollute your water with the carcinogen vinyl chloride and the suspected carcinogen DEHP. You should be aware too that many contaminants in the ground can pass directly *through the walls* of these plastic pipes into your water. Find out if they are used in your area.

Lead pipes have long been used in some water systems, especially for service connections. In Chicago, lead service connections were actually *required* until recently. Home plumbing systems with lead pipes and even lead solder in newer copper plumbing systems can add significant amounts of highly toxic lead to the water, as they did in Washington, D.C. (see chapter 1). Copper plumbing systems in new homes are actually the *most* likely to contaminate drinking water with lead. Most people see copper piping and think they are safe, not knowing about the danger from lead solder. Pollution is most severe when the water is first run after it has been in the pipes for several hours. Leave the tap on for a minute or so to flush out possible contaminants if the water hasn't been run for awhile (unless you live in an apartment building, where water is normally being used around the clock). It's best to use *cold* water for cooking since hot water more readily absorbs contaminants from your plumbing system.

Forty million Americans currently drink water containing more than the 20 parts per billion lead limit recommended by the EPA as a new standard (the existing standard is 50 ppb), and too many health experts believe even the 20-ppb limit is much too high. The use of lead solder in plumbing systems was banned in 1987, but it goes on polluting the water in the millions of homes where it was used up until that time.

TESTING YOUR WATER

Even if your water has been given a clean bill of health by local authorities, having one or more samples tested is highly recommended. Water that has long been considered above reproach is frequently found to be contaminated, so it's worth having at least a basic test run to assess your water quality. The peace of mind you will gain if all is well will more than offset the cost of the test.

The instructions from the lab(s) you choose will give you specifics on how to sterilize your faucet and fill the sample container. Follow these instructions to the letter. Run the water for a few minutes before taking the sample so you won't use water that has been in the pipes long enough to absorb unrepresentative amounts of contaminants. It's a good idea to test the *hot* water too, because contaminants can accumulate in the water heater. Most people don't realize that water heaters should be flushed out and cleared of sediments regularly, once every six months. If you go away and do not drain or at least turn off your hot water heater prior to leaving, be sure to run it until the water turns cold when you return. This should be done if you don't use the heater for a period even as short as two weeks.

Pollution of your water source *can* be intermittent (different contaminants will often show up at different times), so annual sampling is a good idea. High water in the spring is often the time of maximum water contamination. Drought periods, however, can also be risky since individual contaminants may be more concentrated. Test your water in the spring one year, and near the end of the summer the next year.

WATER PURIFICATION IN YOUR HOME

Installing water-purification equipment *inside the home* is becoming an increasingly popular way of dealing with many problems discussed in this book. Many homeowners have installed such "point-of-use" purification equipment rather than trust the quality of city water. Many are so convinced of the dangers of drinking community water that they forego testing and go directly to this option.

The cost of such equipment ranges from fifty dollars for a water filter that fits on a faucet or showerhead to thousands of dollars for some of the more sophisticated high-capacity purification systems that are on the market. (See the end of this chapter for further information on specific systems you may want to consider).

An increasing number of community water systems are themselves investing in water filters for installation in the homes they serve. They are admitting they supply bad water. Meeting the water-quality requirements of the Safe Drinking Water Act can be an enormous financial strain, expecially for the 38,000 U.S. community water systems that provide water to less than 500 people each. So the installation of water purification devices inside customers' homes can save these water utilities a great deal of money by deferring construction of expensive new treatment facilities, wells, and pipelines.

One of the principal decisions you'll be faced with, should you elect to install water purification equipment, is whether you want to purify *all* the water coming into your home, or just that used for drinking and cooking.

The water with which you fill your spa or swimming pool can be at least as great a health threat as drinking water. Most solvents, many pesticides, trihalomethanes, and radon are among the many toxics that can "get under your skin" and into your blood and tissues if you bathe, swim, or shower in water contaminated with them. They can also get into your lungs after evaporating into the air you breathe.

It has been estimated that a fifty-pound child spending an hour splashing in a pool or spa filled with polluted water absorbs *ten times* the toxic materials than would be taken in by drinking a quart of the same water. Higher water temperatures, breaks in the skin, rashes, and sunburn further increase the rate at which toxics pass into the body. A California study measured the rate of absorption into different parts of the body of the insecticide parathion, which is a known mutagen and teratogen for animals, and causes nervous system damage. The accompanying table summarizes the researchers' findings.

| HUMAN ABSORPTION OF PARATHION ||
Body Area	Percent Absorption
Scalp	32.1
Ear canal	46.5
Forehead	36.3
Forearm	8.6
Palm	11.8
Abdomen	18.4
Scrotum	100.0
Ball of foot	13.5

Although it will cost considerably more than cleaning drinking water only, if you live in an area where the volatile organic chemicals used in agriculture or industry are likely to be encountered, it is a good idea to install a high-capacity filter and treat all of the water to which you are exposed, whether via drinking, cooking, showering, swimming, or soaking.

THE KINDS OF PURIFICATION SYSTEMS

As pointed out earlier, a combination of water-purification techniques must be employed to remove *all* the impurities from water. Filters made of synthetic fibers or activated carbon (or a blend of the two) are the predominant methods of point-of-entry water purification. Although large-volume reverse-osmosis units capable of processing the waste water from an entire manufacturing plant are common in industry, their cost has so far prevented household use. Like distillers, domestic reverse-osmosis systems provide only enough water for drinking and cooking.

Filtration is the most commonly used means of cleaning up water. *Activated carbon filters* and *reverse-osmosis filters* are predominant. A variety of *fiber* filters are also used. Some filters employing *powdered* carbon are also still made, but these *should be avoided* since there have been problems with the powder being taken up by the water, complete with the contaminants it has filtered out, and passing into the water that is consumed.

ACTIVATED CARBON

With an activated carbon filter, water passes through either granules or a solid block of carbon. Although the carbon does physically filter out some of the larger impurities that are suspended or dissolved in water, it is effective primarily in removing much smaller, and often more dangerous, impurities. _No other filtration method is as effective as activated carbon in removing many of the extremely toxic organic chemicals often found in water._ Activated carbon removes more of the toxics than the other filters, and does it better.

Toxics are removed as the water passes through the carbon by a process called _adsorption_. Activated carbon is _extremely_ porous— approximately _one pound_ of the material has a surface area of _one acre_. This honeycomb of minute pores attracts and traps pollutants passing through it; the impurities are _adsorbed_. The effectiveness of an activated carbon filter is dependent on several factors:

- The amount of time water is exposed to the carbon (the longer the better);
- The manner in which the water flows through the filter medium (the more diffuse its path, the better);
- The number and range of size of the micropores in the carbon (the greater the range of micropore size, the greater the range of contaminants removed);
- The amount of water the carbon has already filtered.

An activated carbon filter is an excellent choice for a home filtration system—but only if you give it proper maintenance. When an activated carbon filter is installed, it may remove nearly 100 percent of the smaller-molecule synthetic chemicals from processed water, but with time, as more of the micropores are filled, its efficiency will drop. When a filter has been used too long, some of the pollutants it has trapped will be bumped into the water that gets through to the consumer by chemicals with a greater affinity for their resting place.

In an overloaded carbon filter, for instance, molecules of the insecticide aldrin, which is very strongly attracted to carbon, could dislodge molecules of the carcinogen benzene, which has a very low affinity for carbon. The benzene would then be free to pass through into the finished water supply you drink. Lead and other materials that may become lodged in the filter are especially easily displaced. A greater proportion of other pollutants will simply pass through a filter that is overdue for replacement.

The risk that harmful bacteria will become established in the filter medium is also increased by waiting too long for replacement. Filters that use carbon that has been treated to resist bacterial growth are more resistant, but in areas with many bacterial contaminants in the water, even these can become contaminated. So it's a good idea to change filter media at recommended intervals or, better yet, *more often*. Some experts recommend changing them *twice* as often. If your water filter hasn't been used for several hours, running water through it for ten seconds or more will reduce bacteria levels in finished water.

If you use chlorinated city water, a simple test can determine whether your filter is functioning correctly. You can test the water that has been processed for the presence of chlorine with an inexpensive kit from: Hach Chemical Company, PO Box 389, Loveland, Colorado 80539; phone 800-525-5940. If the chlorine is getting through, some of the toxic elements in the water are too.

You can extend the life of a carbon filter by using a less expensive *pre-filter* to remove some of the impurities from the water. Many carbon filters come with a pre-filter to remove sediment. Those with a pre-filter in a *separate* canister are superior to filters with carbon and pre-filter sharing a container, because they are easier and cheaper to maintain. These sediment filters will pick up many of the larger contaminants that would otherwise pass right through the carbon. So this combination is a very good idea.

REVERSE OSMOSIS

Reverse-osmosis filters force water through a plastic membrane, leaving the majority of pollutants behind. Most of the particulates, radioactive materials, and dissolved solids that might pass through a carbon filter are removed. In that respect, reverse osmosis is superior to an activated carbon filter, but it misses many of the very small toxics that are so prevalent. Radon is among the toxics that can pass through a reverse-osmosis filter.

A good combination filtration system will consist of *three* filtration steps performed in three separate containers. A particulate filter strains out the sediment; next the reverse-osmosis filter removes some of the smaller dissolved materials and the larger organic chemicals; the activated carbon then filters out the volatile organic chemicals and other very small contaminants. The pre-filter in this sort of system will typically have to be replaced every few months, while the reverse-osmosis membrane and the carbon filter will last a few years between replacements.

The filter membranes can be the weak link in reverse-osmosis filtration. Some of the thinner membranes can't be used with chlorinated water (without a carbon pre-filter to remove it), while other membranes will become choked with bacterial growth if they are used with *un*chlorinated water. The quality of the filter membrane material varies. Operating at *peak* efficiency, however, a reverse-osmosis membrane will catch 90 percent of the toxics in the water that passes through it.

Small capacity is another disadvantage of reverse-osmosis filters. Most produce only enough water for drinking and cooking—a few gallons per day.

DISTILLATION

Distillation has long been used to purify water. Distilled water is used in scientific experiments because of its exceptionally high degree of purity. Water passing through a typical distiller is heated to boiling, causing it to evaporate and then recondense as pure water, leaving pollutants behind—basically the same process that gives the earth fresh rainwater.

There have been some problems with volatile compounds evaporating when the water is boiled inside a distiller and passing with the steam into the area where the water vapor is recondensed. Some of the newer distillers have overcome this problem. A distiller used in combination with a carbon filter can produce some of the purest water attainable.

The small capacity of most distillers, usually producing only a few gallons per day, and the inconvenience of using many units are drawbacks. Some, although not all, distillers must be loaded by hand and turned on and off manually. All require periodic cleaning to remove build-ups of scale. Distillers are also expensive to operate—it takes about a dollar's worth of electricity to produce five gallons of water. Distillers with housings made of stainless steel have been found to contribute aluminum to processed water. Distillers are sometimes used in homes with filtered water to supply ultra-pure drinking water.

OTHER TREATMENT METHODS

Several other methods are used to purify water, although one drawback or another has prevented their wide adoption:

Aeration can remove most of the volatile compounds from contaminated water. Simply exposing water to air, in a thin film or a spray, will vaporize many of the harder-to-remove contaminants. Chloroform, radon, and many gasoline components are among the volatiles that will be mostly removed through aeration. Aeration is widely used in large water treatment plants to take solvents such as TCE and other volatiles out of groundwater, and some units small enough for domestic use may soon become available.

De-ionization is the process used in water softeners. *Ion exchange* is relied on to replace the calcium and magnesium that make the water hard with sodium, resulting in water that has lost its hardness while becoming more salty. Ion-exchange water purifiers using hydrogen as the exchange medium, rather than sodium, are used to produce pure water for laboratories, but are not yet available for use in the home.

Ultraviolet radiation kills most of the harmful bacteria in water. One drawback of this process is that it can be costly to run the quartz-mercury vapor lamp that emits the purifying ultraviolet radiation.

Now that you have had a good look at the quality of your water and at some of the principal tools available to clean it up, you're ready for *action*. See the reviews at the end of this chapter for more details on water filters, their cost, their installation, and the chemicals removed by each.

BEWARE OF FRAUDULENT FILTERING

Although many reputable companies sell water-purification equipment, unscrupulous companies and salespeople are increasingly attempting to cash in on people's fears about polluted water to make a quick dollar. To keep from becoming the victim of a less-than-honest salesman you should avoid:

- Companies that won't divulge details about their purification system through the mail or over the phone, but insist on sending out a salesman;
- Door-to-door salespeople who want you to sign up for expensive filtration equipment on the spot;
- Anyone who appears at your door saying they are from the water department (or the health department, the state water quality agency, etc.) and wants you to buy water filtration equipment (ask for identification—bona fide water officials will have it);
- Claims that devices are maintenance-free—all water purification devices will require at least some maintenance.

Many water-filter sources offer *free testing* for some toxics (Sears stores being one) and this can be a way to learn more about the quality of your water. It's best to remember, however, that such companies have a product to sell and that there *have* been instances where test results were falsified to promote a sale.

In California, a law making it illegal to "disseminate a false advertisement of a water treatment device by any means for the purposes of inducing the purchase, rental or lease of water equipment devices" went into effect in 1987. The use of false or misleading claims about the performance of a water-purification device or about the health effects of pollutants found in drinking water is also prohibited in California. Similar action has been taken or is being considered in other states. Of course, there are still those who will violate these laws. But such regulations are a step in the right direction.

WHAT ABOUT THE MINERALS IN WATER?

You'll hear claims from the purveyors of certain water purification systems saying that their product is superior because it *doesn't* remove minerals essential to health (carbon filter salesperson), or that it is superior because it *does* remove minerals (distiller or reverse-osmosis filter salesperson). What's the truth?

Since many of our diets are deficient in essential minerals such as iron, chromium, selenium, zinc, cobalt, and manganese that are often present in drinking water, it makes sense that leaving the minerals in the water can be beneficial to health. In addition, many water drinkers say removing *everything*, including minerals, from drinking water results in a "flat" taste. Some studies have shown a higher rate of heart disease and cancer in parts of the country with soft water, which is low in trace-mineral content.

Those who favor *removing* the minerals from drinking water argue that water supplies an insignificant fraction of our total intake of minerals, and that it's better to drink really pure water. They might further argue that unhealthful levels of sodium are often part of mineralized water.

One solution that might satisfy both schools of thought is, luckily, available. Mineral supplements can be added to water that has been ultra-purified. "Dehydrated sea water" and similar products are available at many natural food stores. Or you may simply take a daily vitamin/mineral supplement orally.

VITAMINS AND MINERALS CAN HELP FIGHT TOXIC EFFECTS

A concoction you could wash down with a glass of water (preferably the *filtered* variety) to protect you from toxic substances in the air and water with no serious side-effects—*that* would be a popular item. But such products are already commonly available. They are vitamins and minerals. Whether they come from food and drink or from a capsule, such health aids can protect the body from the effects of the toxics to which we all are invariably exposed.

Preventing the build-up of toxics inside the body is the surest way to avoid negative health effects. You can minimize your exposure to toxics carried into your homes through the water system by taking the steps suggested in this book. But contamination comes in many forms, including pesticide residues in food, air pollution (both outdoors and inside the house), and exposure in the workplace, so an additional measure of protection is desirable.

Luckily, our bodies have a built-in system that provides remarkably effective protection from toxic intruders: the immune system. Maintaining the health of our immune system is one of the keys to preventing the intrusion of harmful elements. Vitamins A, C, and E and the minerals selenium and zinc can help the immune system eliminate toxics from the bloodstream before they become lodged in the body.

Another way of reducing the effects of the toxic elements that find their way into the bloodstream is to put out the "no vacancy" sign at the places where they are most likely to lodge—fatty tissue and bones. Calcium, selenium, and vitamin D play essential roles in the maintenance of healthy, contaminant-free bones.

A healthy diet, adequate exercise, and regular visits to the family physician are the most important ways of protecting your body from the effects of the toxics that have become an inescapable part of daily life.

A physical examination can uncover potentially serious disorders before they become catastrophic.

Regular, nutritionally correct meals help maintain the chemical balance within the body. A greater percentage of the pathogens present in the body are eliminated when healthful food is present. (Moreover, most of the basic requirements for vitamins and minerals will be supplied by the food consumed.) When the body is low on fuel, more toxics are taken up. Body fat is tapped for extra energy, and the bones supply extra minerals. Bones and fat are two of the leading places where toxics such as lead accumulate, however, so these harmful elements can also be released into the bloodstream. Reducing fats in your diet can reduce your body's capacity to absorb toxics.

Working up a sweat on a regular basis helps purge toxic elements (while burning off some excess fat). Several studies have linked regular exercise with a lower rate of absorption of toxics.

Smoking cigarettes drastically increases the body's vulnerability to toxics in the environment. Vitamin C, which plays an important role in the regulation of toxics in the body, is destroyed. Cadmium, nickel, and other potentially harmful elements are found in cigarette smoke, and air-borne pollutants can be converted into a more dangerous form when drawn through a burning cigarette into a smoker's lungs.

While a balanced diet will supply the body with the majority of the minerals and vitamins needed to maintain health, many researchers have shown that taking certain supplements at levels considerably larger than those normally recommended can help the body fight off toxics. However, an excess of some vitamins and minerals (such as vitamin D and selenium) can be harmful. Consult your physician and read up on nutrition before you decide what dosage is appropriate for you.

Heavy metals and many toxics are especially dangerous because they accumulate in living tissue. Once established in bones and nerve cells, metals such as cadmium, lead, and mercury and the more long-lived toxic chemicals tend to stay there. If exposure continues, their concentration increases. Serious disorders such as those discussed in chapters 4 and 6 can result.

The following vitamins and minerals have been shown to help the body to fend off the effects of toxic substances:

- *Vitamin A:* Promotes the growth of healthy cells that are more resistant to the effects of toxics and enhances the health of the body's immune system.

● *Vitamin C:* Reduces the absorption into the body of harmful elements and neutralizes some toxics already lodged in the body. The formation in the stomach of nitrosamines and similar toxic compounds from the nitrates in food is slowed by vitamin C.

● *Vitamin D:* Helps the body metabolize calcium and phosphorus and plays an important role in the building and maintenance of bones.

● *Vitamin E:* Helps maintain a healthy immune system. Vitamin E also has been shown to prevent the absorption of lead, to reduce the effects of cadmium (with the help of selenium), and to control the damage done by mercury and silver. It promotes the growth of healthy cells more capable of fighting off toxic invaders. Research has shown that vitamin E can reduce the effects of carbon tetrachloride, benzene, ozone, nitrous oxide, nitrates, and nitrosamines.

● *Calcium:* Intercepts lead and other toxics in the bloodstream and carries them to the bones and fatty tissue, where they will do less damage than they would in vital organs. A calcium deficiency causes the body to take the mineral from the bones. Toxic elements stored in the bones can then make their way into the bloodstream as a result.

● *Iron:* Reduces the passage of lead through the walls of the intestines into the bloodstream. Some experiments have shown that iron may work with vitamin C to eliminate cadmium from the system.

● *Phosphorus:* Helps remove lead from the bloodstream.

● *Selenium:* Traces of the mineral, although harmful at very high doses, help promote a healthy immune system. Some studies have shown that the health-damaging effects of cadmium and mercury are reduced in those taking selenium supplements. Toxic cadmium and selenium compete for lodging sites within the body, so a selenium deficiency can accelerate the build-up of cadmium.

● *Zinc:* Like vitamin C, zinc can intercept some toxics while they are in the bloodstream and eliminate others already stored in the body. Zinc is believed by some researchers to slow the absorp-

tion of lead and other toxics from the intestines into the bloodstream.

WATER PURIFICATION EQUIPMENT

There are several factors you'll want to consider when deciding
which treatment device (or combination of devices) to buy:
- How much does the system cost?
- What guarantees are offered?
- What is the device's capacity—should *all* the water in your
  house be filtered, or just drinking water?
- What pollutants must be removed, and how pure do you want
  the finished water to be?
- How hard is it to install and use the purification device?
- How much will maintenance cost, and how often must it be
  done?
- Are components (such as plastic) that may taint finished water
  used?

CLASSES OF CONTAMINANTS

No water-purification device is capable, by itself, of removing *all* of
the toxics from drinking water. However, we can get a clearer idea
of just what pollutants *are* removed by the principal domestic
water-treatment techniques by reviewing the three broad classes of
contaminants:

- *Microorganisms*—These microscopic plants and animals include bacteria, viruses, protozoans, algae, and cysts. At least some
such minute life forms are found in most water, and they are the
most common source of problems with drinking water quality.
Chlorine, ozone, and other disinfectants are added at the water
company's treatment plant to control microorganisms, but not all
are removed, and the distribution system and your household
plumbing can add more.

Reverse osmosis is an effective way of removing all but the
smallest microorganisms. Activated carbon filters, on the other

hand, can actually become a breeding place for bacterial growth, which can eventually build to the point that water quality is compromised, although no instances of individuals being infected by growths in their water filters have been recorded. Silver is incorporated in some carbon filters to discourage such growth, but there is no guarantee that even these filters won't eventually become the home of potentially harmful organisms. Many of the larger microorganisms can easily pass through an activated carbon filter. The sellers of some activated carbon filters recommend that their product not be used to filter water with a high bacterial content.

Distillation kills most microorganisms when the water being processed is boiled, and the rest are left behind when the water is vaporized. But bacterial growths can occur in the tank holding processed water (whether that water is purified in a distiller or a reverse-osmosis system). Periodic cleaning and treatment with a disinfectant can minimize the likelihood of bacterial growths getting started.

● *Particulates*—Small pieces of materials that don't dissolve in water are simply taken for a ride between individual water molecules. Radioactive materials, arsenic, asbestos, selenium, barium, lead, mercury, cadmium, cobalt, chromium, and manganese are among the toxics that may be so suspended in your drinking water. Rust, dirt, sand, and other sediments find their way into your drinking water in the same manner.

Reverse osmosis, distillation, and fiber filters are the best ways of removing particulates. Carbon filters allow most suspended material to pass through. The sediment that does become lodged in a carbon filter can impair its effectiveness in removing volatile chemicals. If your water contains both volatile industrial chemicals and heavy metals such as lead, there is a danger that heavy metals will accumulate in a carbon filter only to be dislodged later when water with a higher concentration of volatiles enters.

● *Dissolved solids*—Potentially toxic materials such as fluoride, nitrates, sulfates, and salts are water-soluble. When exposed to water they are chemically altered—dissolved—to become a part of the water. Reverse osmosis, distillation, and the appropriate

fiber filter are the best ways of removing dissolved materials from
your drinking water.

● *Volatile chemicals*—All non-particulate substances that can
be vaporized are volatiles. Chloroform and other trihalomethanes,
most pesticides, radon, PCBs, solvents such as TCE and TCA,
petroleum components including benzene and xylene, and the
majority of the chemicals used in industry are among the volatile
compounds commonly encountered in drinking water. The EPA has
identified 700 volatile chemicals that have been found in drinking
water supplies.

A reverse-osmosis filter will remove the larger molecules of
volatile chemicals, but smaller molecules can pass through. Some
volatile chemicals that are evaporated in a water distiller can get
into finished water (see the review, later in this chapter, of the
distiller made by Scientific Glass for a description of an innovative
way around this problem). An activated carbon filter is the most
effective way of removing volatiles from drinking water.

ACTIVATED CARBON FILTERS

The Egyptians and other ancient cultures used carbon (in the form
of charcoal) 4,000 years ago to clarify and remove odors from their
drinking water, so its use is far from new. *Activated* carbon, however, is a recent development. First manufactured for use in gas
masks during World War I, it is only during the last twenty years
that activated carbon has been used to remove impurities from
drinking water. During the 1960s, a handful of community water
systems used primarily powdered activated carbon to remove objectionable tastes and odors from drinking water. Only during the
last decade have carbon filters made their appearance in home
water systems.

As mentioned earlier, it's best to steer clear of water filters that
employ *powdered* carbon, since the tiny particles of carbon—along
with the toxic materials they have picked up from the water already
filtered—sometimes pass through into finished water.

Carbon that has been molded into a *block* can be superior to
filters relying on a *granular* activated carbon. More carbon is

compressed into a given space in such a filter, encouraging a more diffused flow pattern that puts water into more intimate contact with a greater number of pores in the carbon matrix. However, if the filter is properly designed (and maintained), carbon in *either* block or granular form can be effective.

The *amount* of carbon in the filter (a filter with more carbon won't have to be replaced as frequently) and the *range* of size of the micropores in the carbon (a wider range of pore sizes will accommodate a wider variety of toxic lodgers) are also important considerations when deciding which filter to purchase. "Activated alumina" is added to some filters to remove pollutants such as fluorides and arsenic that aren't intercepted by activated carbon. The aluminum is chemically altered to trap the pollutants, and added to the filter media. Although the evenness of waterflow through a filter is also crucial, an even flow can be accomplished in filters employing either block or granular carbon. Methylene chloride, a toxic chemical used to clean the carbon that goes into an activated carbon filter, has been found in the finished water from some filters.

The following reviews aren't intended to be exhaustive, but rather will give you an idea of the *kinds* of GAC filters available today. Contact the companies offering products that may fit your needs, and check the local phone book for the many firms that doubtless sell water filters in your area. The recommended interval between filter cartridge changes is listed for each filter, but this figure will vary depending on the number of contaminants in the water.

Brita America, Incorporated
321 Commercial Avenue
Palisades Park, NJ 07650
The Brita Water Filter System is a fast-selling unit (about thirty dollars in most markets) that looks like a pitcher. You pour water directly from the tap into a reservoir at the top of the 1.75-liter pitcher. The water passes from there through a small filter that contains activated carbon and an ion-exchange resin. Filters are replaced approximately monthly, and cost about five to seven

dollars. *The Times of London*, in tests of several filtration systems, concluded: "Our all-around favourite—the Brita was actually the cheapest and best buy. . . .Removed every trace of E. Coli bacteria, plus 97 percent lead, 95 percent copper, and all the chlorine." Brita headquarters are in West Germany.

Coast Filtration
142 Viking Avenue
Brea, CA 92621
714-990-4602

The company's twin-canister, point-of-use carbon-block filter, the Water-Safe TOC-200, features a reusable sediment pre-filter. The hard plastic housings on the filter are designed to fit under the sink. One gallon of water per minute is processed and delivered via polyethylene tubing to a gooseneck faucet that mounts by the sink. A total of sixteen ounces of carbon (of two different types to broaden the spectrum of chemicals removed) is used. The sediment filter must be washed periodically and replaced every one to three years (at a cost of fifteen dollars). The TOC-200 costs $190. It is recommended that the carbon block be replaced annually (at a cost of approximately twenty-five dollars).

Coast Filtration also makes a point-of-entry filter, the Concept 2000. The filter uses fifty-two pounds of granular activated carbon. It can treat ten gallons of water per minute. The carbon in the filter is covered by a five-year warranty. A new filter elements costs about $800 (the activated carbon in the filter can be cleaned for about $400). The Concept 2000 water filter's suggested list price is $5,000 (although some dealers sell the system for as little as half this amount).

Cuno, Incorporated
400 Research Parkway
Meridian, CT 06450
203-237-5541

Cuno's AP600 point-of-entry filter uses almost six pounds of granular activated carbon (in a plastic cartridge mounted inside a stainless steel housing). The unit delivers eight gallons of water per minute. The manufacturer recommends changing the filter cartridge every six months (a standard replacement cartridge costs about fifty-seven dollars; a heavy-duty replacement element costs approximately sixty-four dollars.) The suggested retail price of the AP600 filter is about $530.

Environmental Purification Systems
PO Box 191
Concord, CA 94522
415-682-7231 (800-874-9830 in California)

A pound and three-quarters of granular carbon enclosed in a stainless steel housing make the EPS filter one of the best point-of-use filters. Water enters through the bottom of the filter and is slowly pushed through the carbon to the top, thereby eliminating the tendency for channeling common to many filters. The filter is designed for under-sink installation. The manufacturer recommends that the carbon in the filter be replaced every six months. The EPS Drinking Water Filter costs about $350 and a replacement filter costs about ten dollars.

General Ecology
151 Sheree Blvd.
Lionville, PA 19353
215-363-7900

The company's Sparklepure point-of-entry filter relies on a blend of activated carbons and other materials combined in a "structured matrix" to remove a wide spectrum of pollutants from water. According to the manufacturer, the even flow-pattern means the filter is several times more efficient than granular carbon because water is forced through several times as much filter media, eliminating channelized water flow. The filter's housing is stainless steel with brass fittings. Its capacity is twenty-six gallons per minute. The manufacturer recommends that the filter cartridge be replaced every six months (at an approximate cost of $175). The filter medium is wrapped in a plastic mesh with plastic caps. The Sparklepure filter costs about $834. It is not recommended for use with bacteria-rich water sources.

The company's Seagull IV point-of-use filters use materials similar to those employed in the Sparklepure, but are smaller. Suggested retail prices range from about $240 for the Model X-1B (which mounts on the cold water supply line under the sink and provides one gallon per minute of water) to about $400 for the Model X-2F (which has twice as much filter medium as the X-1B). The X-1B will filter 1,000 gallons of water before carbon replacement (new carbon costs about forty-five dollars), and the X-2F will filter 2,000 gallons before replacement (for about eighty dollars). The model X-1F, which costs approximately $300 is designed to be installed by homeowners who feel they can tackle their own installation.

Nigra Shower Filter
5699 Kanan
Agoura, CA 91301
818-889-6877

The company's water filter is designed to attach to a showerhead to remove contaminants from running hot water (it is more difficult to remove volatile chemicals from hot water than from cold). More than twenty ounces of granular activated carbon in a plastic case removes chlorine, trihalomethanes, and many organic chemicals from shower water. The filter cartridge must be replaced every two to four months (at a cost of close to twelve dollars). A flow-restrictor comes with the shower-conversion package, and the filter won't perform as efficiently if it isn't used. The Nigra Shower Filter costs about forty dollars.

Multi-Pure
9200 Deering Ave.
Chatsworth, CA 91311
818-341-7377

Multi-Pure's Model 500 point-of-use filters consist of an outer fiber pre-filter (made of cotton and cellulose) and an inner filter composed of a blend of activated carbon, cellulose, and polyethylene, all enclosed in a stainless steel canister. The company recommends that the filter cartridge be changed after 500 gallons of water have been filtered. The Model 500 costs about $300 and a replacement filter costs about thirty dollars. It will clean one gallon of water per minute (but the flow rate declines with use).

Sears, Roebuck and Company

Sears offers a low-cost way to filter the water in your home. The company's point-of-entry filter consists of a two-stage sediment filter for removing particulates and an activated carbon filter, for about thirty dollars. Sears recommends that a replacement cartridge, which costs a little under twenty dollars, be installed after 800 gallons of water have passed through the filter (cartridges for the sediment filter cost close to eight dollars for the coarse element and about nine dollars for the fine, and replacement is recommended after 3,000 gallons of water have been filtered).

Sears' point-of-use filters use three filtration stages (two particulate filters and activated carbon) to purify enough water for drinking and cooking. The filter costs about ten dollars, and filter cartridges, which are recommended for replacement every three months, cost about four dollars. Sears offers a free water test to those interested in buying a filter.

DISTILLERS

Distillation produces water that is virtually contaminant-free. Microorganisms and most toxics are removed, but, as pointed out earlier, some volatiles can vaporize and find their way into finished water. A distiller (or a reverse-osmosis system) installed on the kitchen sink to provide high-quality drinking water in a home with a point-of-entry filter can be extremely effective.

Durastill
PO Box 76641
Atlanta, GA 30328
816-454-5260

Durastill's water distillers range from the Model 30H (which is manually operated and capable of producing eight gallons of pure water per day, with a suggested retail price of a little under $400) to the Model 46C (an automatic unit capable of distilling twelve gallons of water per day, with a suggested cost of about $530). Models with holding tanks for processed water and portable cabinets cost a few hundred dollars more. A vent for volatile gases that may vaporize in the boiling chamber is provided, and water is run through a carbon post-filter on all models to remove any remaining volatiles. Automatic models feature carbon filters both before and after the distiller. The manufacturer recommends that outlet filters be replaced every six months. The company's distillers are made of stainless steel.

Polar Bear Water Distillers
829 Lynnhaven Pkwy., Suite 119
Virginia Beach, VA 23452
800-222-7188 or 800-523-6388
Polar Bear distillers feature a stainless steel housing and a carbon filter that removes chemicals that find their way into finished water. The company recommends that a new cartridge be installed in the post-filter every six months (a new filter costs about eight dollars). Prices start at approximately $350 (for a manual unit capable of producing eight gallons per day). Eight- and twelve-gallon-per-day versions are available.

Scientific Glass
113 Phoenix, NW
Albuquerque, NM 87107
800-841-9803 (505-345-7321 in New Mexico)
The working parts of the company's Rain Crystal 8 Water Distiller are housed in a Pyrex glass case. In a unique process, water entering the unit is first slowly heated to about 200 degrees Fahrenheit as it passes through a coiled tube surrounded by steam. Volatile chemicals are converted to a gas in the process, and are removed from the water. The pre-treated water is then vaporized and recondensed, as is done in most distillers. A granular activated carbon filter (housed in a Pyrex glass cylinder) then removes any remaining impurities. The distiller comes in a Plexiglas plastic case. It produces eight gallons of purified water per day, and costs $649. A replacement carbon filter costs about ten dollars.

REVERSE-OSMOSIS FILTERS

Reverse osmosis filters are an effective way of removing inorganic elements such as heavy metals and radioactive materials from drinking water. The concentrations of some of the larger volatile chemicals that can most easily pass through an activated carbon filter are also greatly reduced when the water is forced through a reverse-osmosis membrane.

Three basic kinds of reverse-osmosis membranes are available: *cellulose acetate* and the newer, more efficient *polyamides* and *thin-film composites*. The thicker cellulose membrane, when it is new, will intercept more than 90 percent of the fluoride present in source water. However, less than 40 percent of the arsenic and less than 10 percent of the trichloroethylene are removed.

Polyamides and thin-film composites are considerably more efficient. Water passing through a new, properly functioning thin-film filter will lose all but a few percent of its fluoride content, almost 70 percent of its arsenic, and 78 percent of its TCE. Many reverse-osmosis systems employ fiber and/or activated carbon filters to improve the purity of the water produced. One disadvantage of a reverse-osmosis system is its inefficiency—only about 10 percent of the source water that passes into a typical system passes through the membrane.

The following are representative reverse-osmosis systems:

Aquathin
2800 W. Cypress Creek Blvd.
Ft. Lauderdale, FL 33309
305-772-0343
Aquathin manufactures a full line of reverse-osmosis filters with an activated carbon post-filter to further cleanse water after it has passed through the purifying membrane. The company's system uses de-ionization (see discussion earlier in this chapter) to remove contaminants (such as nitrates) missed by the reverse-osmosis membrane. The Model PSS-1 features a cellulose tri-acetate membrane (for use with chlorinated water) packaged in a spiral-wound cartridge. Its retail price ranges from about $700 to $850. Processed water is stored in a four-gallon tank under the sink, and delivered via a gooseneck faucet that mounts on the kitchen sink.

The Aquathin MegaChar point-of-entry GAC water filter is a point-of-entry GAC filter (with twenty-five pounds of carbon medium) that will purify water at flow rates of up to twelve gallons per minute. The system automatically flushes out accumulated sediments every three days to prolong the effectiveness of the filter media. The manufacturer says the activated carbon needs replacing just once every two years, even in areas with especially high concentrations of contaminants. Replacement carbon costs about $200.

Water Resources International, Incorporated
2330 W. Sherman
Phoenix, AZ 85009
602-257-0510

The WRI system produces five gallons per day of clean drinking water. Sediment is first removed with a pre-filter. A reverse-osmosis membrane (both cellulose and thin-film composite membranes are available) then removes smaller particulates and part of the toxic chemicals. A carbon-block filter then takes out the remaining toxics. The company recommends that the sediment pre-filter and the carbon post-filter be replaced every six months (the pre-filter costs about eight dollars, and the post-filter costs about sixteen). It is recommended that the reverse-osmosis membrane be replaced every two years (at a cost of $70 to $120). Retail prices for the WRI system start at about $4,000.

INSTALLING YOUR WATER-PURIFICATION SYSTEM

If you can muster sufficient plumbing skills (either your own or those of a bribeable friend!), you can significantly reduce the cost of a water-purification system by installing it yourself. Before you undertake such a job, however, be sure the warranty of the device you have purchased won't be voided by homeowner installation.

The difficulty of installing water-purification equipment varies. Most point-of-use devices are relatively easy to install. The most basic (and least effective) filters simply attach to the faucet, either on the kitchen sink or shower. Countertop distillers and reverse-osmosis systems, the least expensive models available, purify water from the kitchen faucet without under-the-counter plumbing alterations. Point-of-use systems that mount under the counter and dispense water through a separate spigot above the sink may require complete rearrangement of the plumbing under the sink. Don't attempt the task unless the plumbing involved in installing a new faucet in the kitchen sink would be within your abilities.

Installing a point-of-entry water filter can be a little more difficult, depending on the size and location of the supply pipe and the materials of which it is made, but even this job shouldn't prove insurmountable to many home handypersons. If you would tackle the plumbing for a new shower in the bathroom or for an addition to your home, you should have no problems with installing even the most complicated point-of-entry system. Some rerouting of the main water pipes servicing your home and some remodeling to facilitate filter installation and maintenance may be necessary.

Whether or not you elect to install your own purification equipment, it is important to remember that regular maintenance must be performed on any water-purification device. Filter media and reverse-osmosis membranes must be replaced, storage tanks must be drained and cleaned, and the accumulated scale must periodically be removed from a distiller. Make the job easier by installing the equipment where access will be easy.

To minimize the chances of bacteria growth in your water-purification system, all parts should be disinfected (with Clorox or an equivalent) during installation and as a routine part of maintenance. If you're using copper sweat-solder fittings, be sure to use the new low-lead solders.

MINIMIZING WATER USE

Minimizing the amount of water that will be used in and around your home can make the water cleanup job easier. Water-conserving toilets and showerheads that are designed to work efficiently while using about one-third the water required by their conventional counterparts are available from many fixture suppliers.

If you're planning a major remodeling project or a new house, installing a mini-use domestic water system can further trim water use, thereby reducing the water cleanup task. Energy use is also significantly lowered, because the loss of heat from hot water standing in tanks and pipes is reduced. Such water systems make use of electronic controls located at each faucet to deliver water of exactly the desired temperature and flow through a single pipe to the fixture (most home plumbing systems have separate hot and cold water piping systems using half-inch and larger pipe—mini-use systems employ a single three-eighths inch pipe). Watering can also be programmed with some mini-use systems. Contact Ultraflo Corporation, PO Box 2294, Sandusky, Ohio 44870, for a brochure describing one such system.

Another important way to reduce the total amount of water that must be filtered is to minimize the water used *outside* the home. Choosing plants that can subsist with minimal watering for landscaping while limiting grass lawns, installing a drip-irrigation system to provide water for lawns, shrubs, and the garden (an alternative to sprinklers), and even using the "gray" water that drains out of the kitchen sink or the bathtub for watering—these measures can all improve the efficiency and prolong the life of a whole-house filtration system.

INDIVIDUALS WHO TOOK ACTION TO CLEAN UP THEIR WATER

The following case histories are composites drawn from hundreds of actual situations in which individuals, in different situations, have taken action to clean up the otherwise hazardous water they and their families were consuming.

The Reynoldses had been living in their new, upscale community in the suburbs of a major California metropolitan area for only a few months when they first heard the frightening rumors. A scientist at a nearby university, it seemed, had conducted a study and had concluded that the cancer rate in the Reynoldses' community was far greater than it should be. The scientist was also claiming that unsafe levels of certain chemical solvents and pesticides were being found in the water supply there.

These rumors became even more alarming when a local activist paper finally published an investigative report on the scientist's claims and confirmed that water tests it had paid for revealed what it called "hazardous" levels of chemical solvents and pesticides with tongue-twisting names. The newspaper said these chemicals had been shown to be carcinogenic, mutagenic, and (in one instance) capable of causing male sterility at levels just a little higher than those detected.

The Reynoldses spoke with a neighbor who worked in city government, who told them that the newspaper that had reported the story was noted for "rabble-rousing," and that the community's water supply was among the best in California. Unimpressed, the Reynoldses called the city water department and were immediately referred to the city health department. An official there told them that the community water supply "meets all current health requirements." The official would neither confirm nor deny that there were traces of solvents and pesticides in the water.

After speaking with the scientist whose report had occasioned their alarm, the Reynoldses were convinced that their water did, indeed, pose a potentially serious hazard to themselves and their children. They eventually helped to form a citizens' group that is presently putting pressure on city government to better regulate the suspected industrial sources of the groundwater pollution in their area.

In the meantime, the Reynoldses and many others in their community have taken *individual* steps to safeguard their water. The Reynoldses consulted the experts in water safety in their own area (finding some by simply looking in their Yellow Pages under such headings as Water—Testing), sent away a sample of water to be tested, and tapped into some of the national information sources listed in the Resource Section of this book.

They called, for example, the 800 number of the National Pesticides Telecommunications Network to get specific information on the pesticides identified in their water. They got in touch with the Citizen's Clearinghouse for Hazardous Waste in Arlington, Virginia, an organization started by a mother whose son was poisoned by toxic wastes from the Love Canal. The Clearinghouse provides a wealth of information on toxics and their health effects. It also has a referral list that can put you in touch with toxics experts throughout the country.

And from the Water Quality Association (again see Resources), they got information on the various home water-purification options. After looking over all the information they received, they settled on a system that effectively cleaned up their personal water supply. Because of the potential extreme hazard of the chemicals that had been detected in their water—chemicals that continued to flow through their taps while the political/legal battle to clean up the community water system itself simmered and sometimes raged—the Reynoldses wanted a system that would provide them with the purest possible water.

They chose a distillation system capable of distilling twelve gallons of highly pure water daily. They found this adequate for themselves, their six-year-old son, and three-year-old daughter. The system was also equipped with a carbon filter to ensure quality. The system cost them under $600, and they find it easy to maintain. (A lower-capacity system, costing considerably less, would be adequate for most families of four. Three or four gallons of finished drinking water will suffice in most cases.)

Though they were confident their new system was working properly, the Reynoldses—"just to be sure"—had water that had been processed by the distiller tested. Virtually no trace of the solvents and pesticides previously detected could be found.

Celeste was living in a large apartment building on the East Coast, and was concerned about the quality of her water for the simple reason that it was "rusty-looking." It was an older building, and Celeste worried that the pipes might be lead. She spoke with the building superintendent. He told her that minor corrosion in some of the building's iron pipes was responsible for the red tinge in the water. He said no lead pipes were used inside the building, but that he thought the pipes that came from the water main *might* be lead. He also said that new buildings sometimes have more lead content in their water than do the old ones, since lead is often used to connect joints of copper pipe.

Celeste still worried about the "rust" in her water and wondered if lead might not still be a problem even if, as the superintendent indicated, only the service lines leading to the building were of the old lead variety. She called the city water and health departments and was told quite frankly that, yes, some of the service lines into the older buildings were made of lead. These were gradually—as it turned out *very* gradually—being replaced. Celeste was told to consult private testing firms listed in the Yellow Pages to get her water checked.

She called three companies and found that she could have her water tested for lead and several other contaminants for between $25 and $100. She selected one of the companies and delivered a water sample to the laboratory. The results confirmed her fears. There was lead in her water in concentrations just slightly above those considered safe. Actually, the lab said some would consider the results "borderline."

Celeste alerted her neighbors in the building to the problem and began looking for information on ways to filter her water. At the library she found an article in a national magazine that was helpful in some ways but discouraging in others. She hoped to find a purification method that would be easily portable and inexpensive. She knew there were some carbon filters available that cost relatively little, but the article stated that these are "not effective in removing metals like lead." The other systems talked about in the article cost hundreds of dollars.

Celeste decided she would buy "spring water" at the grocery store, but was quickly daunted by the accumulating costs and the inconvenience of lugging water home every day or two. In addition, she disliked the slightly "plastic taste" some of the water had and worried about its purity, too.

At a party she ran into an old friend who had a background in engineering. She discussed the problem with him. He told her that the magazine was mistaken, that some carbon filters *do* remove lead from water. He said that, for a time, manufacturers of these filters were reluctant to make this claim for fear of provoking regulatory agencies which might assert that definitive tests hadn't yet been done.

Celeste's friend told her about a couple of very inexpensive filters using activated carbon and other materials that *would* remove lead (and other metals) from her drinking water. She selected a thirty-dollar countertop model that removed most of the lead and all of the chlorine from her water. Silver was mixed in with the carbon to discourage bacterial growth. She had to replace the filter element about once a month (a replacement cost six dollars). Her water was no longer discolored, tasted better than ever before, and was almost entirely lead-free.

The Gallaghers lived on forty acres about forty miles from a medium-sized Southern city. They, like all their close neighbors, had wells. There were times when they and their children thought their water tasted a little funny, but they usually attributed these taste changes to "minerals" in the water.

But then one day the Gallaghers' teenage son said the water "tastes like gas." Mrs. Gallagher noticed it next. She said she could sometimes smell a hint of gasoline—the kind used in auto-mobiles—while standing over her kitchen sink. Finally the prob-lem could be ignored no more. One of the Gallaghers' neighbors said he had noticed the same problem, with increasing regularity.

The Gallaghers started getting water in town and transporting it back to their home in gallon jugs. Meanwhile they called the county health department and got the name of a testing lab. The lab found their water to be contaminated with significant levels of gasoline *and* bacteria. They were told that the bacteria were prob-ably coming from a septic tank in the area. And a buried gasoline storage tank that had sprung a leak was the most likely source of the gasoline.

The Gallaghers themselves did not have any gasoline tanks on their land, but they soon learned that a neighbor three properties away did. This neighbor, however, said there was no odor of gas in *his* water supply, and he was clearly reluctant to drain and abandon his underground tank. It wasn't until more than a year later, with talk of litigation in the air and more neighbors complaining of "gas odor" and taste in their water, that this individual had the gas siphoned out of the tank and the tank dug up. He did not reveal whether there were detectable leaks, and the gas problem has persisted in several water supplies. If the tank was the source, the problem wouldn't necessarily disappear once it was removed. Traces of the gas may continue to be detected for years.

The Gallaghers consulted various water-purification companies and decided to use a reverse-osmosis system in their home. (Reverse osmosis is better than carbon filtration when it comes to removing bacteria.) The system they selected included a carbon filter that has been effective in ridding their water of its service station aroma.

While the reverse-osmosis system effectively cleaned up the Gallaghers' drinking water, they still occasionally noticed a gasoline odor when taking a shower, and Mrs. Gallagher had recently begun to notice a gasoline odor in the laundry room after a load had finished. A local store had run an ad in the home improvement section of the Sunday paper, and the Gallaghers decided to stop by and look at one of the carbon filters mentioned in the ad that would filter all the water coming into their house. They finally decided to purchase the filter, and after Mr. Gallagher installed it the gasoline odors disappeared.

9 · The ABCs of Community-Wide Water Cleanup

*T*aking care of the source of your water is a task that requires cooperation. Alone, you're unlikely to be able to improve the quality of your water at its source. As an individual, all you can do under most circumstances is to monitor, and to some extent—as discussed in the previous chapter—improve the quality of the water *after* it enters your home.

Great expense is involved in most public efforts to protect water sources. New sewage plants and improved municipal and industrial waste dumps, all very costly, are needed to reduce the volume of contaminants entering these sources. Reducing the volume and toxicity of industrial wastes by converting to more efficient processes can cost even more.

A total of $12 billion was spent on the construction and operation of municipal sewage treatment plants in 1984, when 170 million Americans were served by sewers. An additional $7.5 billion was spent on industrial waste water treatment plants. The treatment of drinking water cost about $6 billion during 1985, and the expenditure is expected to reach $8 billion by 1995.

But even though these enormous sums have been spent on

purifying water, the overall volume of pollutants from sewage plants has remained relatively constant, and the number of dangerous man-made chemicals in many water sources is increasing. While the waste water flowing out of the average sewage plant or industrial treatment facility is of better quality than it was a few decades ago, this improvement has been offset by dramatic increases in both the volume of total waste and the variety of toxics it contains. The cost of improving all the nation's sewage-treatment plants to the point that they don't cause significant pollution of lakes and streams has been estimated to be around $100 billion.

It can be difficult (and expensive) to do anything about the contamination of a water source, even when it's clear what pollutants are present and where they are coming from. The neighbors in Woburn spent more than $2 million proving in court who was responsible for polluting their drinking water (see chapter 1), and the case was settled before entering the potentially more difficult phase in which they would have attempted to prove the solvents in their drinking water caused the diseases suffered by their families. The geologic studies and consultants' fee required to document the pollution were especially expensive.

As awareness of the threat to health posed by bad water has grown, so have efforts to do something about the problem at the community level. A web of laws and programs administered by a variety of agencies has been woven around the drinking water of communities in an attempt to protect its purity, although, as we've seen, this protection has too often been too little and too late. During the mid-1980s, several federal laws that protect water quality were strengthened, and legislation stricter than federal statutes was passed in many states.

This chapter will describe the principal federal laws that can help protect drinking water and some of the techniques used to clean up contaminated aquifers. See chapter 11 for a description of some of the methods used by various communities (and individuals) to clean up their drinking water.

THE SAFE DRINKING WATER ACT

The Safe Drinking Water Act became law in 1974. A study released that year linked the consumption of New Orleans drinking water with an increased incidence of cancer of the digestive tract. Congress passed the legislation, which had been introduced more than four years earlier, in the midst of widespread public concern caused by the New Orleans study and by the ensuing discovery of dangerous chemicals in the drinking water of communities throughout the country. Cleaning up the nation's drinking water was considered at the time to be a task that would take only a few years to accomplish.

The Safe Drinking Water Act is the only federal law that directly protects the quality of our drinking water. The Clean Water Act, Superfund, and a dozen other federal statutes impact on water quality to one degree or another, but their focus is broader.

The Safe Drinking Water Act was renewed in 1986 after *three years* of debate in Congress. Growing public pressure to do something to improve the still-deteriorating quality of the nation's drinking water once again succeeded in getting the legislation extended, although there were still strong forces who wanted it (and still want it) abolished.

Congress responded to the lackluster performance of the EPA in pursuing the goals of the original legislation by *requiring* the agency to establish standards for a total of eighty-three chemicals by 1989 (standards had been set for a total of only *twenty-two* out of the *more than 700* toxics that had been found in drinking water in the *twelve years* since the drinking water act was first passed). By 1991, standards must be set for another twenty-five contaminants.

There are two kinds of standards—primary and secondary—that apply to drinking water under the Safe Drinking Water Act. *Primary standards* govern toxic elements that are regulated because they pose a potential health threat; *secondary standards* apply to materials that may give an undesirable taste or odor to the water, stain laundry and fixtures (as do some iron compounds), or have some other undesirable effect, although they are not necessarily bad for your health.

The EPA must establish the maximum safe concentration of substances regulated under the Safe Drinking Water Act. A Maximum Contaminant Level (MCL), and a Maximum Contaminant Level Goal (MCLG) are established for each contaminant. The MCL is the maximum level allowed in water supplied by community water systems. The EPA considers the difficulty and expense of compliance when establishing the MCL for a substance. The MCLG is the ideal for drinking water. For instance, the MCL for benzene is five parts per billion, while its MCLG is zero.

In addition to giving the EPA definite quotas and deadlines for its administration of the program, the 1986 Safe Drinking Water Act included a variety of other new wrinkles:

- Required each state to develop a program by 1989 to safeguard the recharge areas of aquifers used by community water systems. (Special protection is available for aquifers that are a community's only source of drinking water).

- Called for more comprehensive management of watersheds that supply drinking water.

- Required certain factories, rural schools, and restaurants that have a self-contained water system that furnishes water to at least twenty-five people to meet the Act's requirements, although they are not defined as community water systems.

- Banned the use of lead pipe and lead solder in water-supply systems and plumbing systems.

- Made it a federal crime to intentionally introduce a contaminant into or to otherwise tamper with a public water system with the intent of harming its users.

- Tightened the regulation of wastes pumped into the ground through injection wells.

- Established new standards for the treatment of drinking water; disinfection, for example, will be required for all community water systems by the end of 1990. About 3,000 of the 9,800 community water systems in the nation that use unfiltered surface water will be required to filter it by 1991, unless an exemption is given. The filtration is intended to reduce the threat to public health posed by water-borne pathogens such as giardia, viruses, and bacteria.

- Established granular activated carbon filtration as the standard to which other water treatment methods will be compared for the filtration of synthetic organic chemicals.
- Allowed the use of "point-of-entry" water purification devices, those that purify *all* the water entering the home, as a way for water utilities to meet water-purity standards. "Point-of-use devices," those that purify only the water coming out of a single tap, *can't* be used to meet the standards.
- Required monitoring for contaminants that are not now regulated. The results of this monitoring will help determine which toxics should be limited in public water supplies in the future. The EPA has published a list of fifty-one organic chemicals for which community water systems must test at least once every five years.

Many water utilities and some state water quality agencies oppose the filtration and testing requirements in the Act, citing the high cost and uncertain benefits. An activated carbon system large enough to filter all the water used in a city of 50,000 costs around $3 million. A plant adequate to filter all the water flowing through a large urban water system costs billions. Administrators of community water systems in lightly populated regions object to being forced to test for most of the same contaminants as a water system in a highly industrialized area.

Although the sticker price for filtration is higher in a town, the *per-capita* cost of filtration is *much* higher for a smaller water system, where adequate funds for operation and maintenance often are not available even without adding an expensive new filtration system. Utilities serving less than 500 customers are allowed to apply for an indefinite number of exemptions from the Act's requirements, as long as they can show that they are improving the quality of their water system. Money is allocated to help the smaller water systems meet monitoring requirements.

The cost of complying with the Safe Drinking Water Act for a typical customer is estimated to range from an average of four dollars per month in a large community water system to between seventeen and thirty-two dollars per month for a system serving less than 500 people. Congress provided funds to help offset the higher cost for those buying water from one of the smaller systems.

Consumers are generally willing to pay for good water. A 1985 survey by the American Water Works Association found that more than half those polled would prefer to pay rates high enough to provide clean water rather than continue to use polluted water from their cities' utilities.

The first phase of the new regulations went into effect during 1987. Maximum concentrations in drinking water were set for eight volatile organic compounds, mostly industrial solvents. In 1988, limits were set for pesticides, herbicides and other synthetic organic chemicals, heavy metals, and disease-carrying organisms found in drinking water. In 1990, the EPA will publish standards for radioactive materials (including radon gas), and a variety of industrial chemicals and pesticides.

U.S. Senator David F. Durenberger (R-Minnesota), chairman of the Senate's Toxic Substances and Environmental Oversight subcommittee, commented on the monitoring program in an article in the *EPA Journal:*

"Over the next two or three years, hundreds of small towns will be surprised to learn that their drinking water wells have been

---

The Need for Wellhead Protection

Currently there are 187,000 public drinking water well systems in the United States. Potential contaminant sources include the following:

- 23 million septic tank systems;
- 9,000 municipal landfills;
- 190,000 surface impoundments;
- 280 million acres of cropland treated with pesticides annually;
- 50 million tons of fertilizer applied to crops and lawns annually (the most common source of nitrate contamination);
- 10 million tons of dry salt applied to highways every winter;
- 2 million gallons of liquid salt applied to highways every winter.

SOURCE: EPA (May 1987)

---

contaminated by unpronounceable chemicals that they had never been warned about. Armed for the first time with adequate information, these communities will, without heavy-handed federal regulation, take the steps necessary to protect their drinking water supplies."

THE CLEAN WATER ACT

The Clean Water Act was passed in 1972 with the goal of making all the nation's lakes and rivers clean enough for fishing and swimming. It prohibited industries and cities from dumping wastes into surface water without a permit. In order to receive a permit, polluters had to agree to meet government standards for waste water treatment.

Congress unanimously passed a renewal of the Clean Water Act in 1987, overriding President Reagan's veto of the legislation. Of the $20 billion authorized to clean up the nation's surface water, $18 billion will help communities build sewage treatment plants over a seven-year period. The renewal authorized an additional $2 billion to pay for programs designed to improve the quality of surface water.

The renewal of the Clean Water Act also tightened the regulations under which licensed polluters discharge wastes into the nation's lakes and streams, and imposed controls on the flow of water from storm sewers and farms into surface waters. The EPA estimates it will cost industrial firms $500 million a year to comply with the new regulations, and projects that as many as sixty-one U.S. chemical plants may be forced to close because of the cost of compliance (putting 3,300 employees out of work). An estimated 23.6 million pounds of toxic pollutants will be prevented from entering water supplies annually as a result of the new regulations.

THE SUPERFUND

The Comprehensive Environmental Response, Compensation, and Liability Act (CERCLA), commonly called the "Superfund," became law in 1980. The $1.6-billion program was established to clean up the nation's chemical waste dumps. The program has since been expanded to clean up other kinds of toxic waste sites. A five- to ten-year effort that would restore hundreds of waste sites was envisioned by Congress in the original legislation.

In 1986, Congress passed a five-year, $9-billion renewal of CERCLA. President Reagan, facing an almost certain override, threatened to veto the legislation, but signed it at the last minute in response to bipartisan appeals.

The "Hazardous Substances Superfund" received $8.5 billion of the total allocated by Congress (the other $500 million went to combat leaking underground gasoline storage tanks). The Superfund allows the cleanup of a waste site to begin before the origin of the toxic substances involved is known. The EPA is given authority to later sue, for the *entire* cost of cleaning up, *any* company that can be shown to have contributed to the problem. Even then, estimates project that less than 4 percent of the money put into the Superfund will be recovered through such lawsuits. If a company found to be liable for cleanup costs can later prove that others contributed to the problem, it can, in turn, recover part of its expenses by suing them.

The five-fold increase in CERCLA funding was an attempt by Congress to keep pace with the bad news about dangerous chemicals escaping from the nation's toxic waste dumps to poison water (and air). With estimates of the total number of dumps potentially requiring cleanup now in the tens of thousands rather than the hundreds, and the estimated bill for that cleanup in the hundreds of billions, it is clear that even this increase in funding is inadequate.

One of President Reagan's principal objections to the new Superfund law was its method of raising money to pay for the cleanup program. A tax on petroleum provides about 30 percent of the total, and a tax on raw chemicals kicks in another 16 percent. But it was the estimated $2.5 billion to be raised through a broad-based tax on corporate earnings that President Reagan disliked.

The CERCLA renewal went beyond the original legislation in several ways:

- The right to sue waste-dumpers was expanded. Citizens can sue in federal court for any violation of CERCLA's provisions.
- The statutes of limitations imposed by the states would start to run only after those suffering a disease discovered that it may have resulted from exposure to toxics, rather than starting at the time of exposure, as had been the case.
- Polluters are required to make information about the toxic materials that are used at their facility available to the public, and to notify local officials of where potentially toxic materials are disposed of or spilled.
- The public's right and ability to participate in the selection of a cleanup method was increased.
- The EPA was required to initiate cleanups at 375 sites by 1991.
- When a Superfund site pollutes drinking water, the EPA is required to supply clean water for all household uses, rather than just furnishing purified drinking and cooking water.
- Funding was increased for studies on the effects of toxics on human health.
- The federal government and the Department of Defense are required to speed up the cleanup of toxic waste sites on their property.
- Standards were set for just how clean a site must be before it can be considered legally "clean." Permanent solutions such as the on-site detoxification of wastes were given preference over the removal of toxics to another landfill that could itself later cause contamination. There is an ongoing argument about what the goal should be for cleaning up water near a Superfund site. Some say the MCL for contaminants regulated under the Safe Drinking Water Act should be the goal while others argue that the goal should be the stricter MCLG. The cost of restoring water quality to meet the stricter guidelines is much higher, and complete restoration may sometimes be impossible.

THE RESOURCE CONSERVATION AND RECOVERY ACT

The RCRA is the federal statute governing the handling of hazardous wastes in business and industry. Companies that use any of the more than 400 contaminants now regulated under the act must obtain a permit. The chemical, petroleum, metals, and transportation industries are the primary sources of the wastes. The rest come from approximately 100,000 smaller concerns, including service stations, auto repair shops, dry cleaners, laundromats, construction firms, photographic processors, laboratories, and schools. The roughly 3,000 facilities that either store or neutralize toxic materials are also regulated under RCRA.

LEAKING UNDERGROUND STORAGE TANKS

An estimated 40 percent of all groundwater contamination comes from leaking underground gasoline storage tanks. *Hundreds of thousands* of tanks holding fuels and other toxic substances at service stations and businesses are expected to spring leaks within the next few years. An estimated 5 to 10 percent of the 1.4 million underground tanks, about half of which hold gasoline, are already leaking.

Leaking tanks have been getting more attention as the result of such shocking news. Congress in 1984 passed an amendment to the Resource Conservation and Recovery Act to address the problem. State water quality agencies were given the authority to carry out the legislation's requirements.

The amendment called for a new approach to the construction, installation, and monitoring of underground tanks. Tanks installed *after* the law went into effect were required either to provide corrosion protection or to be constructed of non-corrosive materials. A sump must be provided to catch tank overflow and spills that occur during the filling of the storage tank or the filling of individual vehicles.

Tank owners are required to install leak detectors, to keep accurate records of their tank-monitoring activity, and to immediately report any leaks discovered. Fines of up to $10,000 per day were authorized for failure to comply with the Act's requirements.

Owners of tanks that contained petroleum products or any of the 700 substances defined as hazardous in the Superfund legislation at any time since 1974 were required by the legislation to register the tanks during 1986. The resulting inventory was then analyzed in an attempt to identify and call for replacement of the tanks thought most likely to start leaking soon.

Many gas station operators say the cost of complying with the new regulations may drive them out of business. A new, corrosion-resistant tank costs an average of $10,000, and satisfying monitoring and record-keeping requirements will also be expensive. The EPA estimates compliance will cost $400 million per year over the next ten years.

Tank owners are also required under the Act to provide liability coverage for each of their tanks, but full coverage isn't available. Insurance companies don't like the odds—damages awarded in court to those suffering the effects of a leaking tank are often in the millions—and as a result offer only limited coverage.

"Investigations just to *define* a contamination problem (from a leaking tank) could cost anywhere from $25,000 to $500,000 and take many months to complete," a 1984 report from the congressional Office of Technology Assessment said. Geologic mapping, numerous water tests, and an assessment of other potential sources of pollution in the area must be performed. Some of the techniques used in containing the spread of pollution through an aquifer are discussed in the next section of this chapter.

An estimated 84 percent of the underground storage tanks now in use have single walls made of bare steel, which is structurally strong but prone to corrosion. Such a tank, installed in a corrosive soil, may start to leak in as little as seven years. A wet climate and soils that are acidic, that have a high electrical conductivity, or that contain sulfides contribute to corrosion. Tanks exposed to saltwater will also corrode quickly. An untreated steel tank installed in an arid climate with neutral soils may last up to thirty years before leaking, however. Corrosion-resistant tanks should last more than twenty years even under the most adverse conditions.

Fiberglass tanks have seen increased use in the last decade. About 100,000 of the 2.5 million underground fuel tanks now in use in the nation are made from fiberglass. While fiberglass tanks are supposed to have a life of forty to fifty years, they often leak much sooner. Underground piping and its connection to the tanks (and cracks in the tank caused by improper installation) are the most common sources of leaks. The alcohol in gasohol and other chemical products can also shorten the life of a fiberglass tank by attacking its resins.

Four percent of U.S. underground storage tanks contain chemicals other than gasoline. The new regulations require that such tanks have double walls or be placed in a concrete vault or lined excavation.

Two other laws recently passed by Congress include provisions relating to groundwater:

The 1986 renewal of the Safe Drinking Water Act required that each state within three years develop a plan to protect the quality of aquifers that are a community's sole source of water. Up to 90 percent of the states' costs in developing the plans is to be reimbursed by the EPA.

In addition, the renewal of the Superfund included a $500-million authorization for a program to clean up already-leaking underground storage tanks.

How a Tiny Bit of Leakage Can Pollute Our Drinking Water

One or two drips per second of gasoline seeping out of an underground tank can seriously pollute an aquifer, given enough time. Such a leak will release about one gallon of gas per day, more than 400 gallons in a year. Depending on the nature of the aquifer under the tank, this rate of leakage can pollute the water supply of tens of thousands of people. Such a leak can be hard to detect since losses are so slow they usually go unnoticed.

THE STATES' ROLE IN WATER CLEANUPS

A prime argument *against* strengthening the dozen or so federal laws that relate to water is that they sometimes usurp the states' authority to control the water within their boundaries. With a resource as varied as water, the argument goes, local control is essential, if the cure for bad water quality is to match the symptoms. Federal laws that have anything to do with aquifers and their use are especially unpopular in the West and other areas where the local economy is often dependent on the unfettered use of groundwater.

The tradition of state and local supremacy in water quality matters today coexists with the strengthened national water pollution regulations. As a result, a partnership has evolved, albeit a sometimes *uncomfortable* one, between state, federal, and local agencies with a statutory interest in water quality.

Federal law is the bedrock of water quality law—the *minimum* water quality protection guaranteed to the nation's citizens in laws such as the Safe Drinking Water Act, which must be enforced in *all* the states.

An increasing number of states have gone beyond this bedrock to better manage the quality of water within their boundaries. Arizona, Florida, California, Iowa, and New Jersey are among the states that are doing the best job of protecting groundwater (see Resource Section for details). California's Proposition 65 is one of a growing number of state laws calling for a stricter approach to the regulation of the toxic materials found in drinking water sources.

Groundwater contamination is one of the most serious problems facing regulators today, both because of the lack of information about aquifers and the lack of comprehensive efforts to protect the resource. Wells serving more than two million people living in twenty-two states have been closed because of pollution. Some have been reactivated after purification equipment was installed, while others remain closed.

Congress failed in 1987 to appropriate funds authorized in the 1986 renewal of the Safe Drinking Water Act for the protection of aquifers used by community drinking water systems. Even then, during 1987 and 1988 35 percent of the EPA's budget, an annual total of $1.4 billion, was spent on groundwater protection programs.

Although aquifer contamination is widespread and unabated, at least *plans* for controlling the contamination have been established or are being developed in most states. Authority to carry out groundwater quality programs has been granted in forty-one states, and thirty-three have developed standards for groundwater quality. However, planning to protect the resource and determining its present condition will be in itself an expensive and lengthy process. Cleaning up the nation's groundwater and permanently protecting it will be more difficult still.

The *enforcement* of the Safe Drinking Water Act varies. The states have the authority to grant variances to certain requirements, and to add and delete certain substances from the list of contaminants for which monitoring must be performed. Some states go well beyond the letter of the law, carrying out routine monitoring near known sources of pollution and actively looking for new contamination sources.

WATER CLEANUP METHODS

How can the flow of pollution from a waste dump into a stream or aquifer be stopped? How can groundwater be cleaned up, once it has become contaminated? What should be done with the toxic materials in a leaking waste dump? These are among the many questions whose answers will determine the quality of tomorrow's drinking water. Some of the answers rely on the cutting edge of science. Others employ techniques that have been in use for centuries. See chapter 11 for specifics.

Much has been learned in recent decades about how to go about cleaning up a lake or stream: Reduce the emissions of toxics from point sources; reduce sediments through proper management of farmlands and stream banks; reduce nutrients from farming, municipal sewage, and industrial processes.

Cleaning up an aquifer is more complicated. As the company

that spent $20 million in the Silicon Valley discovered, just *attempting* to stop the spread of solvents from its leaking underground storage tank can be very expensive. That's why insurance companies today offer only limited coverage to the owners of underground storage tanks containing solvents, gasoline, and other toxic materials.

Stopping the spread of pollution through an aquifer can be a difficult and expensive process. If the contamination is too deep, if it covers too wide an area, or if too much water flows through the aquifer, cleanup may be impossible. Restoring water in a severely contaminated aquifer to pristine condition is next to impossible; the best that can be achieved in most cases is to improve the aquifer's water to drinkable condition, and even that goal can be elusive.

The spread of a pollution plume through an aquifer can sometimes be stopped by a physical barrier placed in a trench dug through the water-bearing strata. Concrete, synthetic membranes, clay, wood, and steel have been used. Such a barrier is only effective for water near the surface, and contamination often finds its way around or under it. Where the volume and flow of tainted water is relatively small, pumps can be installed upstream of the barrier to remove groundwater for treatment and reinjection on the clean side of the barrier.

In a relatively slow-moving aquifer where contamination is less extensive, polluted water can be pumped out of the ground from a series of wells placed where they will intercept the drift of contamination. The water is treated and reinjected back into the aquifer.

Treating polluted groundwater without removing it from the ground is an approach to aquifer restoration that holds promise. Microorganisms or chemicals designed to neutralize the toxic elements are injected into the aquifer, where they restore water quality. Such approaches may be capable of treating a large volume of contaminated groundwater at a much lower cost than techniques involving pumping, but they are still in a developmental stage.

Surface waste disposal still poses one of the greatest threats to water quality, especially groundwater. The disposal of toxic wastes in surface reservoirs has been decreasing, mostly because of a history of such impoundments failing and releasing their toxic contents into neighboring water.

That eighty-mile stack of garbage mentioned in chapter 3 as being about the same size as the nation's annual output of garbage is in reality dispersed throughout the country. The landfills in which it resides too frequently pollute aquifers. The 20,000 toxic waste sites on the list for possible Superfund cleanup represent little more than the tip of the waste-disposal iceberg. The majority of places where dumping, legal and illegal, causes water contamination either are not large enough to be considered for cleanup or simply haven't been discovered yet.

An increasing number of Americans have turned to bottled water as an alternative to the uncertain quality of their tap water. While using bottled water for drinking and cooking is better than using contaminated water, the risk of exposure via bathing and showering mean that the bottled water is, at best, a *temporary* solution to our national water situation. And many bottled products themselves carry dangerous contaminants. See the next chapter to learn how to avoid the pitfalls of using bottled water, the "solution" more and more people, unable or unwilling to do anything to clean up either their community or home water systems, are selecting.

10 · Bottled Water

A hot new product has invaded the soft drink section of most supermarkets. *Water* is the new wave in beverages. Brightly labeled bottles filled with anything from fizzy water with a lemon-lime flavor to imported mineral water from world-famous springs are appearing in an ever-increasing volume, along with plastic gallon jugs of bulk drinking water. Sales are soaring. The promotion of bottled water is similar in glitz and intensity to that normally reserved for a new soft drink or a new wine cooler.

The bottled-water industry, which grossed *$1.5 billion* during 1985, is growing at the rate of 15 to 20 percent per year. Coca-Cola, Anheuser Busch, and other mega-companies now have their own bottled-water labels. At least twenty-five brands of sparkling waters are currently marketed nationally. More than 350 companies sell water products somewhere in the United States.

When France's Perrier company first started selling its mineral water in the United States in the mid-1970s, there was little demand for such products. Many thought it was a fad that would soon fade. There were even Perrier jokes. However, many health-conscious Americans, who felt a growing aversion to both artificially sweetened and sugar-filled soda pop and had an equal mistrust of their tap water, proved to be a ready market for the product. Sales of Perrier went from 500,000 cases in 1976 to *ten million* cases in 1986.

The French each drink fourteen-and-a-half gallons of bottled water annually, the world's highest consumption rate. Residents of Belgium, West Germany, and Switzerland are not far behind with an average per capita consumption of twelve gallons. The average American drinks five-and-a-half gallons of bottled water a year— three times the average of a decade ago. And consumption continues to grow.

New Orleans was one of the first places in this country where the use of bottled water became popular, and it still has one of the highest per capita rates of use among U.S. cities. California leads all states in the consumption of bottled water. Some 17 percent of the state's residents pay to have water delivered to their homes. That figure is 30 percent in Los Angeles.

THE KINDS OF BOTTLED WATER

Sparkling water is bubbly as the result of either natural or man-made carbonation—seltzer, club soda, and some bottled mineral waters make your tongue tingle when you take a drink. The sensation is caused by the little bubbles of carbon dioxide you can see floating to the surface when you pour a carbonated product into a glass. Sugar and other sweeteners are added to many sparkling waters, so read the label if you don't want a drink that tastes like pop.

Carbon dioxide is injected into sparkling water products just before they are bottled, even when the carbonation is "natural." According to standards established by the International Bottled Water Association, water products must be drawn from a natural spring and contain no additives to be called natural. Much of the carbon dioxide that builds up in effervescent groundwater escapes when the water is exposed to the atmosphere. So the gas is captured and reinjected just before bottling.

Club soda is usually filtered tap water to which carbon dioxide and mineral salts have been added. *Seltzer* is filtered, carbonated tap water with no added minerals. Sparkling water products cost about five dollars per gallon, about one thousand times more than tap water. *Water* is a negligible part of the cost of water products. About 40 percent of the retail price typically goes for packaging, 30 percent for marketing, 5 percent for labor and operating costs. The remainder is mostly profit.

Mineral water is groundwater, usually taken from a spring. The water absorbs minerals from the materials of which its parent aquifer is made. These include bicarbonates, calcium, chlorides, fluorides, magnesium, potassium, sodium, and sulfates. Some mineral waters are effervescent. California and Florida require that water contain at least 500 parts per million of such minerals before it can be sold as mineral water. By this standard, the drinking waters of Los Angeles and Houston come close to qualifying!

Europeans drink mineral water for its minerals. Tonics from spas and springs have traditionally been consumed in attempts to cure a wide range of maladies. Such draughts still enjoy much wider acceptance overseas than in this country. The mineral content of bottled water products is often touted in advertisements in other countries. Ads for the same products, when sold in the United States, stress what's *not* in the water rather than what *is* in it. Products that have the strong, medicinal flavor favored in Europe are not popular here.

European bottled waters are considerably more expensive in U.S. stores than are domestic products. Unless the buyer accepts the idea that the blend of minerals in a particular brand of bottled water is restorative, there is little reason, beyond snob appeal, to purchase imported products. Their average quality is no better than that of domestic products. Most health experts say that the minerals taken into the body in drinking water don't have a significant effect on health anyway, since most usable minerals enter the body via the food we eat.

The mineral and sparkling waters are just the crest of a growing wave of water products. The gallon plastic jugs in the beverage section at the store contain *still water*—uncarbonated, unadorned aqua, the same liquid found in the office water cooler. It consists of surface or groundwater that has (sometimes) been filtered, distilled, or otherwise cleaned up. This water is usually treated for bacteriological contamination (primarily with ozone). Still water accounts for 90 percent of total bottled water sales. Such bulk water typically costs in the neighborhood of one dollar per gallon.

Home and office delivery of bulk water accounts for 80 percent of total still water sales. A typical delivery service costs fifteen to thirty dollars per month more than the cost of the water.

Spring water, as mentioned above, must come from a spring, a place where groundwater comes to the surface naturally. Spring water that is forced to the surface by pressure within an aquifer may also be labeled as "artesian" water.

Bottlers aren't required to disclose their waters' sources, however, and there are no restrictions on the names a company that markets water products can choose. Products such as those sold by the "Carolina Mountain Spring Water Company," which sells *filtered* well water, for example, could easily be mistaken for pure spring water by the uninformed consumer. Product labels for companies located in towns with "Springs" as part of their name can similarly mislead buyers. If it's pure spring water, it will be clearly described as such on the label.

The reason for the popularity of the term "spring water" is the aura of purity implied. Water drawn from springs is, in fact, some of the best available. *If* the recharge area of the aquifer feeding the springs is in an area of little pollution, the odds of its water being pure are improved. It also helps if groundwater must filter down through many layers of aquifer materials before coming to the surface.

Mountain Valley bottled water comes from such a spring in Hot Springs Valley, Arkansas. The company says geologic tests have shown it took 3,500 years for the fifty gallons per minute of water that bubbles from the spring to penetrate layers of shale, limestone, and sandstone and then rise to the surface through a geologic fault.

The bottlers of Evian spring water estimate it takes fifteen to twenty years for precipitation that falls on the French Alps to filter down to their spring. The beautiful gardens that surround the company's bottling plant never fail to impress visitors. Employees who report to work with a cold are not allowed in the bottling plant; they are engaged as gardeners until their infection clears up.

Sales of *distilled water,* used in the past primarily for filling irons, cleaning contact lenses, and other tasks requiring pure water, are also booming. An increasing number of smart consumers are using distilled water for drinking and cooking because it is cheaper than many bottled water products and its purity is greater than that of some filtered waters.

Sellers of bottled water must conform to a variety of state and federal statutes. No unverifiable claims can be made about the healthfulness of a product. Information on a product's label must be accurate—although, as noted, the use of the word "spring" in the *name* of the water or company can be misleading. Most water products must meet the same purity codes required for drinking water, although *mineral water* is exempt from regulation in most states.

Even with such regulations, much of the advertising for bottled water products is misleading. For instance, Evian's television ads, featuring physically fit young men and women chugging the product after exercising, claim the product "gets into your body faster than other waters" because of its unique balance of calcium and magnesium. Actually, bottled water products rehydrate the body no faster than tap water.

RATING BOTTLED WATERS

The following bottled waters are available either nationwide or in a large part of the country. Data on the mineral content of water products comes from several sources, including *Consumer Reports* magazine (January, 1987). *Consumer Reports* hired a "professional taster" to evaluate the flavor of the water products reviewed.

While taste is the most noticeable characteristic of a product, it is a highly subjective sensation, about which people all must obviously decide for themselves: One person's elixir is another's poison. It is the *contents* of the water product that are most important. Consult your state or local health department to see what requirements bottled water sold in your area must meet. Be sure to ask whether the agency has tested products marketed in your area.

SPARKLING WATERS

A & P Club Soda, A & P Seltzer—Both are relatively salty and leave a bitter aftertaste.

Appolinaris Naturally Carbonated Mineral Water—Made in West Germany, this mineral water leaves a metallic aftertaste, owing to its very high concentration of dissolved minerals (at more than 2,000 parts per million total dissolved solids, one of the highest).

Bel-air Sparkling Mineralized Water—Relatively high salt content—leaves bitter aftertaste. Bottled for Safeway.

Calistoga Sparkling Mineral Water—Musty aroma and taste—traces of arsenic and fluoride detected.

Canada Dry Club Soda and Canada Dry Seltzer—The seltzer is relatively pure and has no objectionable taste, but the club soda, which is the nation's leading seller among bottled waters, has a slightly metallic taste, a relatively high salt content, and a salty aftertaste.

Cragmont Club Soda—Tops in *Consumer Reports'* rating of sparkling waters; no objectionable tastes or hazardous contents detected.

Crystal Geyser Sparkling Mineral Water—Relatively high levels of fluoride were detected in the product in 1985. The U.S. Food and Drug Administration ordered the company to take steps to reduce the concentrations. The product has a slightly metallic taste and a bitter aftertaste.

Gap 10 Natural Artesian Mineral Water—A sour taste, a relatively high salt content, and a bitter aftertaste have been reported.

Golden Crown Seltzer and Golden Crown Club Soda—The seltzer is relatively salty and the club soda fairly sweet.

Lady Lee Sparkling Water—A bitter aftertaste and a relatively high sodium content.

Montclair Sparkling Natural Mineral—The fairly high salt content gives the product a salty aftertaste.

Perrier Naturally Sparkling Mineral Water—France's Perrier has a metallic taste and a bitter aftertaste because of its relatively high mineral content.

Peters Val Naturally Sparkling Mineral Water—This product of West Germany leaves a slightly salty aftertaste.

Poland Spring Natural Carbonated Water—A relatively pure and neutral-tasting product. Owned by Perrier.

Ramlosa Spring Mineral Water—The product has a relatively high salt content.

Seagrams Club Soda and Seagrams Seltzer—The club soda is relatively salty, has a slightly sour taste, and leaves a bitter aftertaste.

San Pellegrino Sparkling Natural Mineral Water—This product of Italy, lightly carbonated with a slightly metallic taste, has been a favorite among Europeans for ninety years.

Saratoga Natural Sparkling Spring Water—Lightly carbonated with a flat, metallic taste.

Schweppes Club Soda and Schweppes Seltzer—Highly carbonated and relatively pure.

White Rock Club Soda and White Rock Seltzer—The seltzer is fairly salty and otherwise quite pure, while the club soda is sour and leaves a bitter, salty aftertaste.

STILL WATERS

Alhambra Drinking Water—Has a noticeable chlorine and plastic taste, but water is of good quality.

Arrowhead Distilled Water and Arrowhead Mountain Spring Water—The Mountain Spring Water has good taste and quality, while the distilled water has a plastic taste and is bitter.

Black Mountain Spring Water—Good taste and quality.

Carolina Mountain Water—Good taste and quality.

Crystal Drinking Water—Good taste and quality.

Deer Park Natural Spring Water—Has a flat, plastic taste.

Evian Natural Spring Water—France's Evian is of good quality.

Great Bear Natural Spring Water, Great Bear Purified Water, and Great Bear Salt and Chemical Free Water—All the products have a slight plastic taste, but are of good quality.

Hinckley and Schmitt Natural Spring Water—Both have a soapy, metallic taste combined with a fairly strong plastic taste.

Lady Lee Drinking Water—Good taste and quality; rated highest among bottled still waters by *Consumer Reports*.

Mountain Spring Water—Good quality, but with a slight plastic taste.

Mountain Valley Water—Musty taste, but of basically good quality.

Ozarka Spring Water—Good taste and quality.

Pathmark Natural Spring Water—Slight plastic taste, but of good quality.

Poland Spring Water—Good taste and quality.

Safeway Drinking Water—Good taste and quality.

Sparkletts Drinking Water and Sparkletts Distilled Water—A flat taste and a fairly strong plastic taste are minuses for the distilled water, but the drinking water is of good taste and quality.

AN ALTERNATIVE TO TAP WATER

Bottled water sales are a barometer of the public's lack of confidence in the quality of the local drinking water supply. During 1985, New Yorkers were especially concerned about their drinking water. Many switched to bottled water after a letter addressed to Mayor Koch in which the writer threatened to spike the city's

drinking water with plutonium was made public in early April. The
letter said the action would be taken if all charges against subway
vigilante Bernard Goetz were not dropped. This demand was not
met.

A year of unusually dry weather had reduced water levels in New
York City's reservoirs in the headwaters of the Delaware River to
about half of normal by mid-summer of 1985. One hundred million
gallons of water per day were being pumped from the Hudson River
to augment the city's dwindling water supply. Water from the
Hudson normally is not used because of its excessive pollution.
Confidence about the healthfulness of city water was further eroded
as a result, causing a corresponding expansion in the bottled water
boom.

In July, the federal Department of Energy started testing New
York's water supply at the request of the city water department, to
see if the sabotage threat had been carried out. The tests revealed
an unusually high concentration of plutonium 239, a radioactive
isotope that causes cancer.

When the results of the tests were announced in mid-August by
Mayor Koch, he stressed that the concentrations found were well
below federal limits and that city water was perfectly safe to use.
He said the traces of plutonium found in the water were from
natural sources, not sabotage. Few New Yorkers were reassured.

By the end of the month, the sales of bottled water were at an all-
time high. Business was so brisk that retail stores and the ware-
houses that stocked them were left high and dry. Supplies of water
products that would normally last months were sold out in a few
weeks. Many of the city's residents said they wanted to buy bottled
water even if their tap water *was* safe, because the incident had
made them realize how vulnerable the water supply was to sabotage, a growing concern all across the country.

BOTTLED WATER AND OUR HEALTH

People buy bottled water primarily to protect their health. Unfortunately, many of the same contaminants found in drinking water

also show up in bottled water. Heavy metals, solvents, trihalomethanes, pesticides, and even traces of radioactive materials have been found. So switching to bottled water can offer little more than a change of poisons, if caution isn't used.

The many water products that make use of water from a well or a community water system may potentially be polluted by any of the hundreds of contaminants that have been found in the nation's drinking water. Although the water in many products has been filtered, some contaminants will get through the filters, which may themselves become clogged if not changed often enough.

The plastic containers that hold many bulk products can give water a "plastic taste." Those made of polyethylene, the cloudy plastic used in gallon milk jugs, are worst, because the material degrades more quickly than other plastics. As it breaks down, some of its components are absorbed into the water.

A thicker, *very clear* plastic must be used to hold sparkling waters, if they are to retain their sparkle. Such containers degrade more slowly than polyethylene. The health threat posed by contaminants that get into water from plastic containers hasn't been clearly established, but some experts say both the chemicals of which the plastic is made and new compounds that form when the plastic's components react with other impurities in the water can be toxic. It is best to use products that come in glass containers if you use bottled water on a regular basis.

Bacteria, algae, and sediment are the most noticeable pollutants and the leading source of consumer complaints about bottled-water product quality. Monitoring the water source for contamination, keeping bottling equipment clean, and treating water are all technical tasks that can be beyond the capabilities of some bottling operations.

A report prepared in 1985 for the California legislature reveals the problems involved in regulating water bottlers and the kinds of complaints that are received from the consumers of bottled water products in that state. The report says California state agencies have cited water bottlers for:

- Not meeting requirements for monitoring their water source or their product;

- Keeping false or inadequate records of water-test results;
- Doctoring water samples with chorine before testing, to conceal biological contamination.

Consumer complaints mentioned in the report include:

- Algae growth inside water bottles;
- Water smelling like gasoline or chemicals;
- Murky water;
- A parasitic disease caused by using the product.

Water bottlers are as anxious as regulators to see fly-by-night operators controlled. Since purity is all the sellers of water products really have to offer, they see the marketing of impure products as a threat to the entire industry.

An increasing number of states are imposing regulations on the bottled-water industry. Representatives of your state water quality agency or the local health department can probably tell you what requirements those selling water products in your area must meet, and may have test results for some products. Call and find out if the brand you are using has been tested or monitored by local authorities. For a listing of some of the principal brands of bottled water, see the Resources section of this book.

But before we get to Resources, let's see—in the following chapter—how some individuals and some communities have cleaned up their water.

11 · Successful Water Cleanups *Are* Possible

THE "CIRCULAR" SOLUTION

*T*he systems that supply drinking water and treat waste water in a modern city are almost always *linear*—requiring that an evergrowing number of water sources be tapped to meet evergrowing demand. Waste water treatment facilities are similarly and continually expanded to keep pace with increases in sewage volume.

In a linear system fresh water passes through the distribution network; a small percentage is consumed and the rest goes "down the drain" for good. Although we tend to think of "sewage" as laden with contaminants it is actually *99.9 percent pure water!* The wastes in waste water are actually only one-thousandth of the total volume of sewage.

Although a growing number of communities are responding to the shortage of new water sources by recycling part of their waste water (see chapter 2), the overall pattern of use remains linear even in these communities. Recycling and conservation programs simply improve the efficiency of linear use.

Natural systems, on the other hand, tend to be *circular*— "wastes" are used as resources. For instance, a tree takes up nutrients from the soil, and as it grows, its leaves and branches fall and enrich the ground underneath it. When the tree dies, it falls to the ground and rots, further building the soil from which it drew its sustenance while living.

Water also is naturally recycled. Water vapor evaporated from the ocean by the sun collects in clouds that bring rain and snow to land. The water feeds lakes, aquifers, and streams, and eventually returns to the sea. Examples of such self-regulating, self-cleansing cycles abound in the natural world.

The linear approach to quenching the nation's thirst for drinking water and to cleaning up its waste water is a primary *cause* of many of the water quality problems we face today. Spawned in an era of seemingly infinite stores of natural resources, the linear concept of continually expanding water delivery networks and waste water disposal facilities to keep up with growing demand has outlived its time.

The growing demand for today's shrinking supplies of potable water and the growing public insistence that water quality be protected indicate that we need to take a new approach to supplying drinking water and processing sewage. Circular water management systems (which are described below), when properly designed and operated, can make effective use of the nutrients and other "wastes" found in waste water. These elements would become pollutants if introduced into a stream. At the same time, such methods create a new source of potable water where none existed before.

Linear solutions to a community's water needs tend to be short-sighted. They too often pit one city against another, whether in competition for a coveted new water source or because treated waste water from one community pollutes the drinking water source of another.

Although effluent discharged into lakes or streams is supposed to meet federal and state guidelines for concentrations of regulated contaminants, violations are common. Unfortunately, instances of raw sewage being discharged into waterways are not unusual. Friction between the residents of the rural areas from which a community's water supply is taken and city government can also increase as water supplies get tighter. Circular management systems tend to clear up many such difficulties.

Linear systems are self-perpetuating. Businesses and individuals that provide the hardware and expertise to build new water-supply systems or conventional sewage treatment plants generally reject circular water management as "dangerous" and "impractical." Officals in a community that has invested billions in a linear system—developing an expanded water supply and installing the latest in sewage treatment equipment—are not likely to give serious consideration to a radical change of approach, especially since it could be interpreted as an admission of earlier error. If the community again runs short of water or if the sewage plant starts acting up, more of the same medicine is the usual prescription.

DISTRIBUTION OF MUNICIPAL SLUDGE BY MANAGEMENT PRACTICE

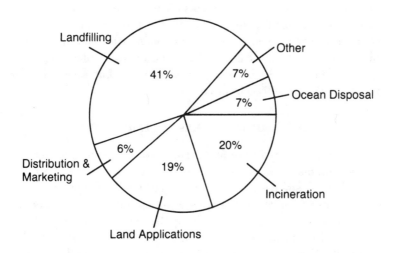

SOURCE: *The Hazardous Waste System* (U.S. EPA Office of Solid Waste Emergency Response, June, 1987).

HOW CIRCULAR WATER MANAGEMENT WORKS

It was estimated in 1980 that if all the wastes discharged into the nation's waters were instead used to grow crops, their value would be *$18 billion*. At the same time, the burden of pollution carried by the nation's waters would be greatly reduced.

As it is, nutrient-rich waste water from sewage plants is a major source of pollution that threatens the health of many lakes and streams. Disposing of the sludge that is produced by a sewage plant also too often taints drinking water sources. The sludge, which is about 80 percent water, is dumped in landfills, burned in incinerators, or dumped in the ocean. In a circular system, by contrast, the "waste" is removed from the water and used as a resource. Clean water is a by-product of the operation. The organic elements in sludge are great soil-builders, and it has been used to revegetate mine tailings, municipal dumps, and waste piles generated by smelters and other industries.

An increasing number of communities are cleaning up waste water with water management that relies more on the type of cleansing processes that work so well in nature and less on conventional (high-tech) water treatment. The processes include natural chemical reactions, the use of microorganisms and plants, and filtration through sands and soils. Comprehensive planning is required to make such circular approaches work, but a host of water problems can be solved in the process.

THE HAMILTON LAKES WATER SYSTEM

A circular system is in use at the Hamilton Lakes development in Itasca, Illinois, four miles west of Chicago's O'Hare Airport. In 1978, Itasca's sewage plant was running at full capacity, and the deep aquifer from which the city's drinking water came was being depleted at an alarming rate. An application had been made to allow the city to use water from Lake Michigan to alleviate the overdraft of the aquifer. It would be years, however, before the water would arrive, even if the request was speedily approved by the state. Flood control was also a problem in the area, and development of the 274-acre site promised to increase an already severe problem, since during a storm, more water normally runs off the surface of developed land.

Years of expensive delay seemed imminent before the development could be built, since neither water nor sewer hookups were available. Plans for a luxury hotel, offices, commercial spaces, and homes would have to be put on hold if a solution wasn't found.

Allan Hamilton, managing partner of the company that wanted to develop the site, proposed an unusual approach to provide drinking water and waste water treatment for the complex. Its water would be pumped from a shallow aquifer that wasn't being over-used, disinfected with chlorine, and treated to reduce its iron content. The water would then pass through the development's distribution system and be used much as it would be in most community water systems. Water-conserving plumbing fixtures such as low-flow shower heads and water-saving toilets would be used throughout the development.

The unusual part would come after the water went down the drain. Sewage would first flow to two oxidation ponds (see sidebar "The ABCs of Oxidation"—later in this chapter), which would be capable of treating a quarter of a million gallons a day. Air would be injected into the waste water during its forty-day stay in the lagoons to speed up decomposition of its nutrients. Conditions favorable to microorganisms capable of breaking down the contaminants in the water would be maintained.

From the pre-treatment lagoons, waste water would be pumped through two sand filters. Finished clean water would be collected in perforated drainage pipes under the filters. This water would be treated with chlorine as a final insurance against biological contamination, and would then be used to irrigate twenty-eight acres of grass, shrubs, and trees on the site.

Part of the irrigation water would soak into the ground and move laterally to feed a series of lakes. Some would penetrate deeper to recharge the shallow aquifer from which the complex's water was to be drawn, thereby elegantly completing a circle of water use, purification, and reuse that would solve the water problems facing the potential development.

The open areas in the planned system would be designed to hold storm runoff long enough to allow most of it to soak into the ground. Water from paved areas and rooftops would be collected and directed onto lawns that would channel rainwater through a meandering path to remove many of its contaminants. Shallow basins would hold up to an inch of rainwater before allowing overflow to pass directly into one of the lakes. The special landscaping would roughly double the amount of water that percolated into the shallow aquifer under the site.

For fire fighting purposes, two of the lakes would provide 3,000 gallons per minute of water for up to eight hours. This water would be pumped through irrigation mains to hydrants located around the development.

The Hamilton Lakes project became a reality in 1982. Five interconnected lakes and the landscaped areas between and around them made the development a pleasing break for the eye in an otherwise built-up area. The water recycling system has been successful here because of the careful planning that went into it. Although the water treatment system appears on the surface to be quite simple, careful engineering—mindful of natural processes—is the reason for its success.

This circular waste-handling system has been even more efficient than had been predicted. Tests conducted soon after it started operation showed that 97 percent of the solids that had been suspended in the waste water and 99 percent of the nutrients had been removed by the time finished water was used for irrigation. Although the state required disinfection of the water before its use for irrigation, the system was doing a very reliable job of removing bacteriological contaminants, even *before* chlorination. During a drought in 1983, the grounds at the complex were green while nearby open areas were parched.

Economics are perhaps the most intriguing aspect of the Hamilton Lakes experiment. Homes and businesses in the development are assessed for water and sewers at the same rates paid by those obtaining the services from other local sources. During 1984, income was more than $20,000 greater than operating expenses, demonstrating that improving water quality can be profitable. In addition, it was estimated that more than $3,000 worth of fertilizer was applied to the development's greenery in the form of nutrient-rich treated waste water.

REMOVING THE WASTE FROM SANTEE'S SEWAGE

Santee, California, a suburb of San Diego, is another community that has escaped many of the escalating uncertainties and expenses surrounding water supply and waste disposal by using circular water management.

An average of just ten inches of rain falls annually in the San Diego River basin, and a sparse supply of water has long plagued Santee. Windmills pumped water from a local aquifer to grow a variety of crops in the valley during the 1940s, when a dam for an upstream community's water supply dried up the lower San Diego River. Farmers' wells soon dried up, as had the river, putting an end to irrigated agriculture in Santee. Subterranean flows from the river had been maintaining water levels within the aquifer.

After World War II, Santee increasingly became a satellite of booming San Diego. Santee joined the Metropolitan Water District of Southern California and received a share of the water delivered through the Colorado River Aqueduct.

Problems with the city's sewage treatment plant, overburdened by an increasing volume of waste water from a growing population, began in the late 1950s. Treated waste water from the plant exceeded state standards for contaminants, and the creek into which the plant's effluent was dumped was being polluted. Santee was urged to join a group of local communities planning the construction of a regional sewer system that would dump untreated sewage in the Pacific Ocean.

The manager of Santee County Water District, Ray Stroyer, didn't agree with the reasoning behind the planned district sewer. He told his board of directors that the city's supply of water from the Colorado River was expensive and certain to become more so, and that he felt it was wrong to divert the precious water across 300 miles of desert, use it once, and then pollute the ocean with it. Stroyer's board agreed to forego the district sewage project and to give him the latitude he needed to look for an alternative way of dealing with the city's sewage.

The solution Stroyer and the board eventually agreed on permanently changed the face of Santee. A string of seven lakes surrounded by grass, shrubs, and trees is now a prominent feature of the town. The lakes were constructed in what had been an abandoned gravel pit, at the edge of town. They are filled with waste water that has been treated using natural processes. The quality of this water has been found to be about the same as the city's drinking water, and is much better than the average output of the conventional treatment plant the city had previously relied on for sewage processing.

Santee's lakes were first filled in the mid-1960s. Shortly after the system began operation, the city qualified for a federal grant to build a sewage treatment plant. Proponents of the new plant believed the city's new sewage recycling system was at best a temporary solution; they said the filtration beds would soon clog and that disease organisms would eventually contaminate the entire system. In fact, however, the oxidation ponds and the soil filtration systems have worked reliably over the years, and travelers now come from around the world to have a look at the Santee system with the idea of building something similar back home.

One of Stroyer's predictions has, in particular, proven to be astute: The cost of Colorado River water and of waste disposal through the district system have both gone up dramatically in the last thirty years. And today, even the *availability* of an adequate supply of water is uncertain. San Diego County has long been "borrowing" water that was found by the Supreme Court in a 1964 decision to belong to Arizona. San Diego County has used 300,000 to 500,000 acre-feet of Colorado River water annually, although it is entitled to only 165,000 acre-feet. City officials estimate that San Diego County water systems could be one-half short of the water needed by the year 2000. Santee, with its already-proven water-purification system producing abundant clean water right in town, is far better equipped than most local communities to face the coming crisis.

In Santee's sewage treatment system, waste water first flows into a sixteen-acre oxidation pond. The sewage stays in the pond for about thirty days, by which time most of its pollutants have been removed. Further treatment is still necessary, however, and *earth* provides it. Several 5,000-square-foot percolation beds covered with dirt are sculpted into the slope of a nearby shallow canyon. Water is pumped from the oxidation ponds into one or another of the basins, where it slowly soaks through the soil. It is collected at the bottom of the filter area and flows into the first of the series of lakes.

The water is purified in a variety of ways as it passes through the soil. First, it is *filtered*. Contaminants in the waste water become lodged in the voids and spaces that exist in all soils, much as occurs in an activated carbon filter.

Chemical reactions also help to purify the water while it is in contact with the soil. Some toxics are rendered harmless when they react with elements in the soil; others become part of the soil; and still others are converted to gases and released into the atmosphere.

Biological processes also remove toxic elements from the water. Countless *microbes* inhabit the top few inches of most soils. They convert organic matter to humus. The humus, in turn, helps in both the filtration and the chemical alteration of contaminants passing through the soil. Certain microbes can also render harmless some of the toxic materials found in the waste water. Others remove nitrogen. Germs, bacteria, and viruses can be neutralized by microorganisms in the soil.

Although they had gone along with Santee's unusual approach to sewage treatment, state officials were far from convinced that the system would produce really clean water. They were especially concerned that viruses would find their way through the system to pollute the water in the lakes. However, after extensive testing, with special attention given to viral organisms, the state approved the water in the lakes for swimming and boating. Lake water is used to irrigate the city's greenbelt, a nearby golf course, and local farms. It also recharges the underlying aquifer.

Soil filtering, such as that used in Santee, is more effective at removing biological contaminants and nutrients than a conventional sewage plant. Most of the nitrogen, phosphorus, and potassium in waste water passes right through such sewage plants virtually untouched. *Billions* of pounds of these nutrients are discharged into U.S. streams and lakes each year from sewage plants, thereby causing excessive algae growth and depleted oxygen.

Soil is, in fact, more *resilient* in handling the constantly changing blends of toxic materials that are typically found in municipal sewage than is a conventional sewage treatment plant. The diverse community of organisms present in soil are better able to *react* to changes in the chemistry of the waste water to be treated. If more are needed to process a substance that is showing up in unusually high concentrations in the sewage, the soil is capable of producing them.

Soil filters are even more effective when plants are grown on them. The plants take up nitrogen and some of the contaminants left in the soil by the waste water, thereby extending the life of the soil filter. Experience shows that crops grown in such a system do not contain chemical pollutants or disease organisms, and they can be sold to offset the cost of operations. Exchanges are becoming more frequent between metro water districts and rural water districts. Farmers sell their water in return for a guarantee that it will be sent back after the city has used it. The farmer can sell his water rights and still have water with which to irrigate his farm. In fact, the nutrient-rich waste water cuts fertilizer costs.

The ABCs of Oxidation

If sewage is emptied into a pond, natural processes will digest many of the nutrients and contaminants it contains. Such oxidation ponds have been used, in one form or another, for centuries in many parts of the world, although just *why* they work has only been known for the last few decades. If properly managed, an oxidation pond can reliably remove most of the contaminants from waste water with less threat to the surrounding environment than that posed by a conventional sewage treatment plant.

In an oxidation pond sewage is broken down by a process similar to the digestion that occurs inside a properly operating septic tank. Wastes from some U.S. military installations were disposed of in oxidation ponds during World War II, and thousands of U.S. communities today use such ponds to pretreat sewage.

In 1924, inadequately treated waste water from the Santa Rosa, California, sewage plant was seriously polluting a local creek, but the community could not afford a new treatment plant. As a temporary solution, city officials decided to close down the plant and pump the city's sewage into a gravel pit.

The sewage did not disappear, as had been hoped. Instead, solids settled out and sealed the bottom of the gravel pit, and the waste water started to fill the pit. Residents of the city braced themselves for a pervasive stench from the unplanned lake, but none occurred. The water that eventually started to flow out of the newly formed pond was as pure as it would have been had it come from the treatment plant that was beyond the city's means.

Since that time, many communities have constructed oxidation ponds, but it was not until the 1950s that an understanding of how they worked was reached. *Oxidation,* the same chemical process that causes a compost pile or a burning piece of wood to get hot, was primarily responsible for the purity of the water flowing out of Santa Rosa's accidental oxidation pond and similar bodies of waste water. Most of the solids in the sewage were "burned"; enough oxygen from the atmosphere was getting into the water to support the slow "combustion" of contaminants in the water. The same process is speeded up in many modern oxidation ponds by injecting air into the pond.

In addition, a variety of microorganisms can remove pollutants

from such a pond. For instance, oxidation promotes the growth of
bacteria that are beneficial to water quality.

CITIZEN ENFORCEMENT OF WATER
QUALITY LAWS

Most of the major environmental legislation enacted in the last
decade has included provisions allowing citizens to initiate lawsuits to force compliance. The idea was to give individuals recourse
if they thought the federal agencies charged with implementing the
laws were not doing their job.

As pointed out earlier, even in the best of times, budget con-
straints and politics can interfere with a public agency's role as a
steward of public resources. In the anti-regulation climate of the
Reagan administration, however, the convenience of the corporations doing the polluting too often received more attention than the
strict enforcement of environmental statutes. As a result, civil
lawsuits are assuming an increasingly crucial role in the enforcement of water quality laws.

The majority of the actions are brought for noncompliance with
the Clean Water Act, the law requiring polluters of surface waters
to acquire a permit and conform to its provisions. Penalties of up to
$25,000 per day can be assessed for violations. The courts have
ruled that it need only be established that the conditions of a
company's permit were violated before fines can be levied. It does
the company no good to contest the validity of the limits imposed in
its permit, or to show that no environmental damage has resulted
from violation.

In 1979, the federal government filed eighty-one cases under the
Clean Water Act. In 1982, after the Reagan administration had
come to Washington, only fourteen suits were filed. Sixteen citizen
actions against violators of the law were filed that year. By 1983, the
pace of prosecution had picked up. A total of seventy-seven suits
were filed by the government, and sixty-two by citizens.

To have legal standing to file such a lawsuit, an individual need only show that he or she uses the waterway being polluted. "Use" can range from drinking its water to using it for recreation. An organization can file a suit on behalf of members who use the lake or stream. The Natural Resources Defense Council and the Environmental Defense Fund are involved in such actions (see Resources Section—Sources of Information).

The potential plaintiff in such a suit must give sixty days' notice of intent to the polluter and to responsible government agencies. If one of these state or federal agencies takes action against the polluter within this time, the citizen's suit must be dropped. It normally takes considerably longer than sixty days for the EPA to prepare a case and the Justice Department to file it, so citizen suits are seldom superseded by action on the federal level.

State governments, however, *are* usually capable of taking action within the sixty-day limit, and when they do, they generally seek lower damages than do citizen plaintiffs. As a result, corporations that find themselves the defendants in an action charging violation of the Clean Water Act, or other water quality statutes, often *ask* the state to sue them to avoid citizen prosecution.

Both government agencies and polluters are taking enforcement of laws such as the Clean Water Act more seriously as the result of citizen lawsuits. Not only must defendants that lose such a suit come into compliance with the conditions of their permit and pay fines for past violations, they must also reimburse the expenses incurred by the plaintiff in bringing the action.

BULL RUN'S CITIZEN STEWARD

The forests that grow on the western slope of the Pacific Northwest's Cascade Mountains are some of the most magnificent in the nation. Giant firs, hemlocks, and cedars with six-foot girths tower a hundred feet above the ground. Dense undergrowth has filled in the space below the canopy of the conifers.

Winter storms that have been gathering energy since their birth in the Gulf of Alaska regularly unleash ferocious gales that blow across the western front of the mountains, the first significant barrier to the wind's eastward progress. The storms also unload much of their moisture as they are forced over the Cascades—more than twenty feet of snow often accumulates in the higher areas during the winter, and *heavy* rainfall can come at any time. It is this ultra-wet climate that makes such prolific growth possible.

This deluge of fresh water would seem to guarantee an abundant, high quality supply of water to the area's community drinking water systems. Although rare droughts do occasionally come along (such as the one during 1987), the western Cascades are one of the most reliably wet areas in the country. But water in many of the streams issuing from the mountains is too often polluted.

Logging is the most frequent cause. Roads crisscross the hills of virtually every watershed along the mountains' western flank, roads that lead to vast clear-cuts. The centuries-old prime-growth trees have almost all already been hauled down these logging roads to the sawmills. The resulting scarcity of the straight-grained lumber they produced has driven prices for products such as "clear fir" through the roof—creating intense demand for the remaining prime-growth trees. Today's clear-cuts are carved out of forests that have been cut at least once in the past. The streams that originate in these logged-over areas invariably carry more silt and other suspended materials that make them less desirable as sources of drinking water. Herbicides and insecticides are also frequently found.

The City of Portland is more fortunate than most communities that draw drinking water from the western Cascades. Most of the city's water comes from the Bull Run River, in the foothills east of town. The river's watershed has been managed for nearly a century with maintaining high water quality as the number one goal. Much of the area from which the river flows is still virgin forest.

President Benjamin Harrison declared 144,000 acres of the federally owned land in and around the Bull Run River off-limits for logging, development, or any public use in 1892 (at the request of Portland city officials, who were looking for better water than that being drawn at the time from the Willamette River). In 1904, Congress passed the Bull Run Trespass Act, which made the area out-of-bounds to all but employees of the water department (or the federal government) engaged in activities that enhanced the river's water quality. The river now provides drinking water for about *one-third* of Oregon's residents.

City officials are proud of their unique water source. In 1984, the Portland City Council appropriated almost $140,000 (which was bolstered by an additional $45,000 in private contributions) to publicize Bull Run water. Six-ounce promotional samples were bottled and distributed around the world. Portland Water Department officials say there is virtually no chance for water coming from the pristine Bull Run watershed to become polluted. They say drinking water flowing from the drainage is as pure as it was when the Bull Run Preserve was created.

Not everyone, however, agrees that the quality of the city's water is beyond reproach. In fact, there are those who say that the city and the U.S. Forest Service have completely botched the management of the Bull Run drainage. They say this mismanagement is the reason that more chlorine has been added continually to the river's water over the years to make it suitable for consumption. Dr. Joseph Miller of Sandy, Oregon, is one such individual.

"The public has been deceived and kept in the dark about the management of the Bull Run Reserve and the quality of its water," says Miller, a retired Portland physician. He is one of the most vital forces of the Bull Run Interest Group, a citizen's organization concerned with management of the preserve.

In 1948, Miller purchased ninety-four acres bordered by the 142,000-acre reserve on three sides. The land was on a ridgetop on the southern boundary of the protected area.

"We felt this was a Garden of Eden," Miller says. "We were very lucky to find land next to a virgin area." The idea that the adjacent land would *remain* untouched and untrammeled by man was especially appealing. "It would have been sacrilegious to go in there," he says. The Miller family frequently drove out to the acreage to enjoy its unspoiled beauty and the wildlife that abounded there.

But an article tucked away on the third page of the *Oregonian* in May, 1971, permanently changed Miller's perception of the Reserve and its management. Titled "Forest Service Proposes Recreation in

Bull Run," the article said that the Forest Service was considering allowing public access to the entire area, even for swimming. The agency was also considering timber sales. The public was given sixty days to comment on the proposed changes.

Miller was shocked. He had believed that the area was permanently protected from such abuses, and from the associated threat to water quality. As a physician, he was far more concerned about the kinds of pathogens that could get into the city's drinking water from human use of the watershed—especially if swimming and fishing were to be allowed—than about contamination of city water caused by animals.

Miller disliked the idea of logging in the reserve, because in addition to the more obvious effects such as erosion, it would bring more people to the Bull Run watershed, thereby increasing the risk of human disease organisms entering the water. He had weathered a giardiasis epidemic in Portland in the 1950s (the suspected source of the giardia cysts was beaver, in the Bull Run watershed, though this was never proven). And though that outbreak had caused discomfort to those who were vulnerable, it was mild compared to bacteria and disease organisms that could enter water from human wastes.

A series of visits to city water department officials and the Forest Service turned Miller's shock to outrage. At the Forest Service office, he got a copy of the timber management plan for the Reserve. The documents disclosed that logging had been going on there without public knowledge, *since 1958*.

Portlanders didn't like the idea of allowing swimming in the city's drinking water supply, and the plan to open the Bull Run drainage was roundly criticized in the public hearing conducted by the Forest Service. As a result, the agency abandoned that portion of the proposed changes, but the agency went ahead with plans to sell more of the Reserve's timber. Local governments supported the plan, since part of the revenues gained from the sales would end up in their coffers.

When it became obvious that yet more timber sales were likely to occur, Miller went looking for a lawyer. The first two he talked to said he would have to *prove* the logging that had already occurred had compromised water quality to have further timber harvest stopped. This would be prohibitively expensive and time-consuming. But the third attorney he consulted said the Forest Service had already broken the Trespass Law of 1904, and agreed to file suit on Miller's behalf to have the logging suspended. The Oregon Environmental Council and the Northwest Environmental Defense Center at Lewis and Clark University in Portland helped with the suit.

The Forest Service defended its action, claiming that logging in the drainage was an essential part of the mandate in the 1904 Trespass Law to protect Bull Run water quality. Fire roads, according to the agency's argument, were essential to prevent the spread of forest fires that could pollute water. But Congress would never authorize money for the construction of these roads, according to the Forest Service, if there were no timber sales to pay for them. Therefore logging was necessary to protect Bull Run water quality.

The plaintiffs countered this argument by producing water quality experts who testified that logging almost always had a deleterious effect on water quality. They also pointed out that logging itself had caused many fires in the Bull Run drainage, with hundreds of acres blackened by slash fires that got out of control. Lightning strikes, in contrast, had caused only a few acres to burn since the Reserve had been in existence.

In March of 1976, the court decided the case in favor of the plaintiffs, saying that the Forest Service had acted illegally in allowing the timber sales. Later that year the agency published an Environmental Impact Statement (which had been requested in the lawsuit) summarizing the management options for the "Bull Run Division," a 100,000-acre reserve set up by the agency in 1959 (see the sidebar for a summary of how the Bull Run Reserve shrank while the city of Portland grew).

The Shrinking Bull Run Reserve

1892—President Harrison sets aside a total of 142,000 acres for Portland's water supply. It includes all the land that drains into the

Bull Run River, all but 4.5 square miles of the watershed of the Little Sandy River, which is located south of the Bull Run drainage, and a buffer area surrounding the two rivers' watersheds.

1959—the U.S. Forest Service creates the 100,000-acre Bull Run *Division.* Buffer zones around the Bull Run drainage are removed from the protected area.

1976—Judge Burns rules the Forest Service's division of the land was illegal. The Forest Service lobbies to have the Trespass Act of 1904 repealed.

1977—Congress repeals the Trespass Act and replaces it with legislation that creates the 95,382-acre Bull Run *Watershed Management Unit,* with the reduced boundaries conforming to a 1976 Environmental Impact Statement prepared by the Forest Service.

1988—A Forest Service Environmental Impact Statement proposes the establishment of an approximately 68,000-acre Bull Run *Watershed Unit,* which will allow extensive logging and public access in most of the drainage of the Little Sandy River, up to the boundaries of the reserved area, and to the edges of the Bull Run watershed.

In 1977, Congress passed a law that replaced the Trespass Act of 1904. The new legislation gave the Forest Service and the Portland Water Bureau the final say in decisions about the management of the city's water source. Public comments were required before changes in the management plan could be taken.

Opponents of the trend toward a smaller reserve with fewer restrictions on its management have testified at such hearings that a return to the original strict trespass laws applying to a larger acreage would be in the best interest of Portland water users.

Miller proposed that the original boundaries of the reserve actually be *increased* to encompass all of the Little Sandy's drainage. He said using water from that river, which has benefited through most of its recent history from the same protection extended to the Bull Run drainage, would be smarter than increasing reliance on well water of questionable quality (the Portland Water

Department's backup wells are located under an eastern Multnomah County district that has tens of thousands of septic tanks and heavy industrial development). But more timber harvesting within a shrinking reserve seems to be the trend for the Bull Run area.

SUING THE EPA TO COMPLY WITH THE
SAFE DRINKING WATER ACT

The Bull Run Interest Group, along with Miller and other individuals and organizations, filed a suit against the Environmental Protection Agency in March, 1988, after that agency failed to meet a deadline established in the Safe Drinking Water Act. The EPA was supposed to have established standards by December, 1987, for deciding which community water systems using surface water sources would be required to filter these supplies. The plaintiffs claimed that the EPA, the U.S. Forest Service, and the Portland Water Bureau have not done an adequate job of protecting Portland's water source. They wanted to see a copy of the filtration rules, and called for a court review of the regulations if they feel these are inadequate.

Miller said the plaintiffs favored a return to stricter measures to protect Bull Run water quality (including a cessation of logging). They feared that if the quarter-billion dollar plant that would be required to filter Bull Run water were built, the Forest Service and the water department would take it as a signal that more of the Bull Run area could be opened to logging (since the filtration plant would remove whatever pollution might result). The plaintiffs contended that if the Bull Run watershed were preserved in its natural state, it would continue to provide water pure enough not to require filtering for generations to come. As of this writing, the Bull Run Interest Group suit was still pending.

Arizona Groundwater Law

Arizona is a state of extremes. Fountains, lakes, and well-irrigated parks and golf courses (complete with water hazards); a swimming pool in every back yard in many neighborhoods; neatly manicured lawns; vast irrigated farms—all surrounded by desert. Such disparities abound. The rugged, *arid* face of the real Arizona shows through just outside the boundaries of the irrigation districts and community water systems that supply the state's residents with the universal cosmetic, water. This water is all that maintains Arizona's fresh, green facade.

Always a dry state, under the surface Arizona is becoming even *drier*. The unrestrained use of water has created the oases that are home to millions. But the sparse precipitation in the southern half of the state, where most of the people live, is not sufficient to recharge streams and aquifers nearly as quickly as they are consumed. Falling groundwater levels and dry streams are the result.

The effects of the long term overuse of the resource are beginning to show. The long wrinkles that reach across the land's surface for miles in areas of subsidence are just the most visible manifestation of the state's shrinking reserves of groundwater. Water is the primary limit to growth in most parts of Arizona.

For decades the federal government has been after Arizona to do something about the over-withdrawal of groundwater. In 1945, the Interior Department said Arizona could receive no federal funds for a precursor of the Central Arizona Project (discussed below) until restrictions were imposed on groundwater use. The state legislature in 1948 passed an act intended to do that, but as the law had no regulatory teeth it had little effect on the spiraling use of groundwater.

When the law proved unworkable, the Arizona Supreme Court issued a decision that established a "rule of reasonable use" to govern groundwater withdrawals. It established a first-come, first-served approach to groundwater use that gave the advantage to the water user with the deepest well and the largest-capacity pump.

The essential nature of water in Arizona has never been a secret. The plants and animals native to the region are, without exception, efficient users of water, since the survival of their species has long depended on reliably finding and hoarding it. But even many of these native species, are finding it increasingly difficult to live in the growing part of the state where pumps have sucked up too much groundwater.

In 1934, Arizona's governor, Benjamin B. Moore, actually called out the state's National Guard to halt construction of Parker Dam on the Colorado River between his state and California. The dam was being built to supply the Colorado Aqueduct system, which would carry water to southern California. Dam construction stopped shortly after the Arizona troops arrived.

The Guard was ordered home when Governor Moore was assured by President Franklin D. Roosevelt and the Interior Department that an irrigation project would be built in the Gila Valley in southwest Arizona. There had been only one Guard fatality—a private who later died of the pneumonia he caught while on the expedition.

Today, with the Central Arizona Project (CAP) already bringing water from the Colorado River to Phoenix, and with completion of the link to Tucson due in a few years, it might seem that Arizona's water problems are at an end. But the cost of CAP water and the uncertainty about its even being *available* during dry years mean that even the CAP is no panacea. How best to use Arizona's scant supply of water will remain a pressing problem for the state.

There is, however, a silver lining to Arizona's overuse of groundwater: The state's aquifers are now protected against overdraft and pollution by the most comprehensive regulations in the nation. The state legislature in 1980 passed a law that limits future municipal and agricultural growth in portions of the state without an adequate water supply. Developers must *guarantee a 100-year supply* of water before residential construction permits will be issued! The goal of this startling but clearly needed legislation is to balance withdrawals of groundwater with the rates of replenishment, and to end the *mining* of aquifers by the year 2025.

The Act presumes that all the state's aquifers potentially may be used to supply drinking water, and should therefore be kept in as good a condition as possible. Operators of facilities that may contaminate groundwater are required, under the law, to obtain a permit. The permit will be issued only if it can be shown that the installation has been designed to prevent aquifer contamination. An impermeable barrier usually must be installed under sites where municipal or industrial waste is dumped before a permit is issued. The law established an in-state "superfund" with $6 million available annually to clean up sites not covered by the federal hazardous waste cleanup program.

A 1977 state law established a commission to address groundwater overuse. Fourteen legislators and eleven representatives selected by the governor were appointed. The law stipulated that if no groundwater legislation was approved by September, 1981, the study commission's recommendations would automatically become law.

It was not the first time such a commission had been set up. It was, in fact, the fourth. The recommendations of the three previous committees, which had been dominated by agricultural interests, had become lost in legislative committees. If the latest group could come to a decision, however, the chance of their findings being acted on were improved because industrial and municipal interests were better represented than they had been in preceding commissions.

The commission's draft report was issued in July, 1979. It recommended curbing the state's overuse of groundwater, and established a goal of ending aquifer overdrafts by 2025. Farmers upset by the draft report showed up at hearings to complain that their right to use groundwater hadn't been adequately protected. Governor Bruce Babbit chaired a marathon series of meetings with commission members over a period of six months, starting in the fall of 1979, to iron out the differences. Secretary of the Interior Cecil Andrus threatened to cut off funds for the CAP if no legislation was passed by the following summer.

Passage of the Groundwater Management Act in June, 1980, forestalled that threat. This new law

- Allowed the regulated sale of groundwater rights, which has turned out to be a financial boon to farmers;
- Allowed water users to continue pumping at 1980 levels;
- Established a tax on water pumps to help fund groundwater-protection programs;

- Imposed restrictions on new water use by industry and mining.

In order to preserve the delicate balance achieved in the governor's meetings, a "non-severability clause" is part of the Act. The clause states that if any of the Act's provisions are found to be unconstitutional, the entire law will be null and void. So far, no successful challenges have been mounted against the legislation.

Although Arizonans still have a long way to go before they achieve the goal of balancing groundwater use with recharge rates by 2025, the state's groundwater law has already had many beneficial effects. Housing developments are now being located where water is available; in the past they were built with the assumption that water would somehow be supplied to them.

Landscaping is also changing. Water hazards on golf courses are being replaced by equally hazardous desert features that do not waste water. Much of the green grass in freeway medians and back yards is being replaced with native plants that require little or no watering. Efforts such as those in Tucson, where water use has been reduced by one-quarter over the last fifteen years as a result of conservation and recycling (see chapter 2), are becoming more common.

Increasingly, more and more communities across the nation are going to have to bite the bullet and match growth with *available water,* or—to put it bluntly—perish.

Resources Section A
Water Information Sources

1. HOTLINES

*A*nswers to questions about toxic chemicals—their use, disposal, and health effects—are available over the phone from a variety of sources, including government agencies, trade associations, and citizens organizations. Here's a list of information sources, with numbers you can dial for help:

THE ENVIRONMENTAL PROTECTION AGENCY

Information concerning federal water quality programs is available to the public and to professionals through several EPA hotlines. You can find out more about Superfund site cleanups and other programs administered in your area by the EPA.

Public Information Center
202-829-3535

If you want to know more about the EPA, its programs or activities, contact the center either by calling or by writing: U.S. EPA, 401 M St., SW, Washington, D.C. 20460.

RCRA/Superfund Hotline
800-424-9346 (202-382-3000 in Washington, D.C.)

Specialists at the center are available (between 8:30 A.M. and 4:30 P.M. EST) to answer queries relating to Superfund and the Resource Conservation and Recovery Act (see chapter 8). You can obtain information about toxic wastes, their handling and disposal, the technical side of the cleanup of leaking toxics dumps, and details on the legal requirements of the laws through the hotline.

Safe Drinking Water Act Hotline
800-426-4791 (202-382-5533 in Washington, D.C.)

The hotline was established to explain the sweeping changes in the 1986 renewal of the Safe Drinking Water Act to the public and to water professionals. EPA specialists on the Act are available between 8:30 A.M. and 4:30 P.M. EST to answer questions about recent developments in the interpretation and enforcement of the law and to help you get access to appropriate documents.

Small Business Hotline
800-368-5888 (703-557-1938)

This hotline was established to help small businesses comply with environmental laws and EPA regulations. It is available between 8:30 A.M. and 5:00 P.M. EST, Monday through Friday.

Inspector General's Whistle-Blower Hotline
800-424-2000 (202-382-4977 in Washington, D.C.)

If you think you know of an instance in which the EPA has been responsible for waste, fraud, abuse, or mismanagement, this is the hotline for you. Complaints are handled confidentially, and even EPA employees and contractors can use it without fear of reprisal. It is staffed between 10:00 A.M. and 5:00 P.M. EST, Monday through Friday.

Toxic-Substance-Control-Act Assistance Information Service
202-554-1404

Callers from the chemical industry, trade and labor organizations, environmental groups, and the general public can find out more about the act and its enforcement.

Chemical Emergency Preparedness Program
800-535-0202 (202-479-2449 in Washington, D.C.)

Chemical emergency planning assistance is available to local governments and others through this hotline (8:30 A.M. to 4:30 P.M. EST, Monday through Friday). *Don't,* however, call this number to report a chemical *emergency.* Dial 911 in that case. See the entry for the National Pesticides Telecommunications Network (below) for emergency information on pesticides.

EPA Regional Offices

Region 1:
JFK Federal Bldg.,
Boston, MA 02203
(617-223-7210)

Region 2:
26 Federal Plaza
New York, NY 10007
(212-264-2525)

Region 3:
6th and Walnut Sts.
Philadelphia, PA 19106
(800-438-2474)

Region 4:
345 Courtland St., NE
Atlanta, GA 30308
(800-241-1754; 800-282-0239
    in Georgia)

Region 5:
230 S. Dearborn
Chicago, IL 60604
(800-621-8431; 800-572-2515
    in Illinois)

Region 6:
1201 Elm St.
Dallas, TX 75270
(214-767-2600)

Region 7:
1735 Baltimore Ave.
Kansas City, MO 64108
(816-374-5493)

Region 8:
1860 Lincoln St.
Denver, CO 80203
(800-525-3022)

Region 9:
215 Fremont St.
San Francisco, CA 94105
(415-556-2320)

Region 10:
1200 Sixth Ave.
Seattle, WA 98101
(206-442-1220)

NATIONAL PESTICIDES TELECOMMUNICATIONS NETWORK

800-858-7378 (806-743-3091 in Texas)

This twenty-four-hour-a-day, seven-day-a-week hotline is a source of general and technical information about pesticides—their use and risks. In cases of suspected human poisonings, data on symptoms and antidotes is available on an emergency basis. In addition to providing data on pesticides' toxicity, the center provides answers to questions about specific products, safety procedures for applicators, environmental effects, and cleanup and disposal practices.

THE CHEMICAL REFERRAL CENTER

800-262-8200

This hotline is operated (from 8:00 A.M. to 9:00 P.M. EST, Monday through Friday) by the Chemical Manufacturers' Association, the principal trade association representing the chemical industry. If you want to know more about the toxicity or handling of a specific chemical, specialists at the center can refer you to the appropriate manufacturer or a regulatory agency.

NATIONAL TOXICOLOGY PROGRAM

919-541-3991 (PO Box 12233, Research Triangle Park, NC 27709)

Information about the health effects of toxic chemicals is available through this program run by the U.S. Department of Health and Human Services.

CANCER INFORMATION SERVICE

The National Cancer Institute has a national center in Washington, D.C., and a network of state offices designed to answer questions about cancer, its causes, prevention, detection, and treatment. Dial 1-800-4-CANCER anywhere in the country to access the network. Phone numbers for state Cancer Information Service centers follow:

Alabama:
 205-934-6644
California
 Northern—213-226-2374
 Southern—213-206-0278
Colorado:
 303-630-5271
Connecticut:
 203-785-6338
District of Columbia:
 202-687-5055
Florida:
 305-548-4821
Hawaii:
 808-524-1234
Illinois:
 312-226-2371
Kentucky:
 606-233-6545
Maryland:
 301-955-8638
Massachusetts:
 617-732-3214
Michigan:
 313-833-0710, Ext. 244
Minnesota:
 612-925-6336

Missouri:
 314-875-3506
New York State:
 716-854-4400
New York City:
 212-207-3540
North Carolina:
 919-286-5515
Ohio:
 614-421-7800
Oklahoma:
 405-455-6261
Pennsylvania:
 215-728-3110
Tennessee:
 615-971-1318
Texas:
 713-792-3245
Utah:
 801-497-2009
Washington:
 206-467-4675
Wisconsin:
 609-263-6919
West Virginia:
 304-293-2370

2. ASSOCIATIONS

A variety of national associations provide information about drinking water and its quality. Manufacturers, government agencies, water quality professionals, and interested citizens have banded together in such groups to set industry standards, to coordinate research, and to publish relevant data for use by members and by the public. The following associations offer services to the public:

American Institute of Hydrology
PO Box 14251
St Paul, MN 55114
612-379-1030
This association of professional hydrologists and hydrogeologists
(who specialize in the flow of groundwater and the structure of
aquifers) establishes certification standards and procedures. A
Registry of Professional Hydrologists and Hydrogeologists is published annually.

American Water Resources Association
5410 Grosvenor Lane, Suite 220
Bethesda, MD 20814
301-493-8600
Engineers, scientists, and businesses interested in water resources are members. A variety of publications are available.

American Water Works Association
6666 W. Quincy Ave.
Denver, CO 80235
303-794-7711
The AWWA's membership is composed primarily of professionals in the drinking water supply and waste water treatment fields. A
wide variety of publications, films, slide shows, and "Waternet," a
computerized water-information database, are available. Although
AWWA information services are primarily geared to full-time water
industry workers, they are also a good source of new information
about standard procedures, as well as new developments in the
supply, treatment, and distribution of drinking water.

Association of State and Interstate Water Pollution Control
     Administrators
444 N. Capital St., NW, Suite 330
Washington, D.C. 20001
202-624-7782
The association is a national professional organization made up
of state administrators of the Clean Water Act. General information
about the quality of the nation's water and progress toward meeting
the goals of the Act are available.

Association of State Drinking Water Administrators
1911 Fort Meyer Dr., Suite 803
Arlington, VA 22209
703-524-2428
Information and publications about the Safe Drinking Water Act
and the states' administration of the Act are available from
ASDWA.

Clean Water Action Project
317 Pennsylvania Ave., SE, Suite 200
Washington, D.C. 20003
202-547-1196
The Project, which started as an outgrowth of Ralph Nader's
water pollution taskforce, is a citizen's group working for strong
water quality laws. Information on drinking water quality and the
health effects of toxics is available.

International Association for Great Lakes Research
c/o Institute of Science and Technology
University of Michigan
Ann Arbor, MI 48109
313-763-1520
Scientist, engineers, and others interested in ongoing research
of the quality of the Great Lakes are members. The association
publishes a quarterly journal and sponsors an annual conference.

International Bottled Water Association
113 N. Henry St.
Alexandria, VA 22314
703-683-5213
The association's members are water-bottling companies, mar-
keters of water products, and equipment manufacturers. Publications relating to bottled water, its marketing, and quality are
available from IBWA.

International Water Resource Association
University of Illinois
208 N. Romine
Urbana, IL 61801
217-333-0536

Individuals interested in the development of water resources are members. Publications include a magazine, books, and conference proceedings.

Lake Erie Cleanup Committee
c/o John Chascsa
3568 Brewster Rd.
Dearborn, MI 48120
(313) 271-8906

This association is composed of fifty-nine citizens' groups from the U.S. states and Canadian provinces surrounding the Great Lakes. Their primary goal is the restoration of the quality of Lake Erie's water. The group is affiliated with the National Wildlife Federation.

Lake Michigan Federation
Eight S. Michigan Ave., Suite 2010
Chicago, IL 60603
312-263-5550

The federation conducts conferences, workshops, litigation, and public education programs relating to Lake Michigan. Organizations, individuals, and educational institutions are members. The group encourages citizen participation in public policy decisions that affect water quality.

National Demonstration Water Project
1111 N. 19th St., Suite 400
Arlington, VA 22209
703-527-2282

NDWP is an alliance of local groups interested in the development of adequate, affordable drinking water supplies and waste water disposal systems in low-income rural areas. Publishes the *Rural Water News*, a bi-monthly publication.

National Water Supply Improvement Association
PO Box 1344
Springfield, VA 22151
703-256-2680

The desalination of sea water and the recycling of waste water are the primary focuses of this group of researchers, businesses, and individuals. Publications include a monthly newsletter and conference proceedings.

National Water Well Association
PO Box 16737
Columbus, OH 43085

Aquifers and the use of groundwater in community drinking water systems are the predominant subject of NWWA publications. The association is made up of water equipment manufacturers and water-supply professionals.

Passaic River Coalition
246 Madisonville Rd.
Basking Ridge, NJ 07920
201-766-7550

The Passaic River, which flows through southern New York and northern New Jersey, supplies drinking water to three-and-a-half million people. The coalition is made up of organizations, foundations, municipalities, and individuals concerned about the river's quality. A variety of publications relating to the river are available.

Water Pollution Control Federation
601 Wythe St.
Alexandria, VA 22314
703-684-2400

The Federation is a nonprofit educational organization concerned with the treatment of municipal and industrial waste water. Local associations that are WPCF members are located in most states, and in five Canadian provinces. The Federation publishes books detailing all phases of waste water treatment, a technical journal, a newsletter, and public information pamphlets summarizing the disposal of sewage and hazardous wastes.

Water Quality Association
4151 Naperville Rd.
Lisle, IL 60532
312-369-1600

This association of water purification equipment dealers and manufacturers sets standards and handles consumer complaints regarding household water-cleanup equipment. Pamphlets describing commonly encountered water quality problems and other water quality information are available.

Water Resources Congress
3800 N. Fairfax Dr., Suite 7
Arlington, VA 22203
202-488-0688

The group, which is composed of state and local government agencies as well as agricultural, industrial, labor, and civic organizations, promotes the improvement and development of the nation's rivers, harbors, and lakes. A variety of publications are available.

3. OTHER INFORMATION SOURCES

Center for Science in the Public Interest
1755 S St., NW
Washington, D.C. 20009
202-332-9110

CSPI conducts research, and provides information about health and the environment to the public.

Citizen's Clearinghouse for Hazardous Waste
PO Box 926
Arlington, VA 22216
703-276-7070

The clearinghouse, founded by a mother whose son was poisoned by toxic wastes in Love Canal, supports groups whose purpose is to stop such pollution in their own neighborhoods. Information on toxic wastes and their health effects—and referrals to experts—are available through the Clearinghouse.

Citizens for a Better Environment
59 E. Van Buren, Suite 1600
Chicago, IL 60605

Research, information, and litigation relating to water quality, sewage-sludge management, and toxics found in drinking water are some of CBE's principal areas of expertise.

Environmental Action
1346 Connecticut Ave., NW, Suite 731
Washington, D.C. 20036
202-833-1845
This national citizens' organization is concerned with protecting environmental quality. Improving water quality by lobbying Congress for more stringent regulations, conducting research into the extent and the effects on health of pollutants, and initiating lawsuits to force compliance with environmental laws are some of the primary means used to achieve this end. The organization's bimonthly magazine, *Environmental Action,* is a good source of up-to-date information on toxic wastes and drinking water quality.

Environmental and Energy Study Institute
410 First St., SE., Suite 200
Washington, D.C. 20003
202-863-1900
The Institute is a non-partisan policy analysis organization that works closely with the Congressional Environmental and Energy Study Conference. Among recent EESI reports, *Groundwater Protection: Emerging Issues and Policy Challenges* is of special interest.

Environmental Defense Fund
257 Park Ave. South
New York, NY 10010
212-505-2100
EDF is a national citizens' organization that has had an immense impact on U.S. environmental law over the last two decades (the battle over the banning of DDT in the early 1970s was the organization's first major victory). EDF staff scientists research the health effects of toxic chemicals, the EDF lawyers initiate litigation to force both polluters and government agencies to comply with federal environmental law. The *EDF Letter,* a monthly newsletter, features articles relating to the health effects and regulation of toxic wastes.

Freshwater Foundation
2500 Shadywood Rd.
Navarre, MN 55392
612-471-8407

The Foundation publishes two monthly newsletters: *U.S. Water News*, which reports new developments relating to water supply, treatment, conservation, and waste water treatment; and *Health and Environment Digest*, which focuses on the health effects of environmental contaminants.

Golden Empire Health Planning Center
2100 21st St.
Sacramento, CA 95818
916-731-5050

The Center offers residents of the Sacramento Valley technical assistance with problems relating to toxic substances in the air and water. It offers an extensive environmental health library, as well as workshops and other presentations relating to the area's water quality.

National Network To Prevent Birth Defects
PO Box 15309, Southeast Stn.
Washington, D.C. 20003
202-543-5450

The Network's newsletter, *Birth Defects Prevention News*, reviews new research relating to the effects on the human fetus of toxics such as lead, nitrates, radioactive materials, industrial wastes, and pesticides. Information about learning disabilities, leukemia, and other health effects of toxics that may affect the developing fetus is available from the center.

National Water Center
PO Box 548
Eureka Springs, AR 72632
501-253-9755

This group, composed of individuals interested in preserving pure water sources through conservation measures, is a source of publications about composting toilets, water-conserving plumbing fixtures, and waste water and drinking water treatment.

Sierra Club
730 Polk St.
San Francisco, CA 94109
415-776-2211

The Sierra Club, the granddaddy of the environmental organizations, is a good source of information about water quality and the contaminants found in the nation's water, and has long supported air and water quality in Congress and the courts. The club's magazine, *Sierra,* reports new developments in water quality.

Water Education Foundation
717 K St., Suite 517
Sacramento, CA 95814
916-444-6240

The supply and quality of California's water are the Foundation's focus. WEF members are water professionals and interested individuals. Publications available include the bi-monthly magazine *Western Water,* which examines new developments and issues relating to California's water; a California water resources map that details the state's natural and man-made water features; and laypersons' guides to California water and its quality.

Water Pollution Control Federation
601 Wythe St.
Alexandria, VA 22314
703-684-2400

The Federation is a non-profit, educational organization concerned with the treatment of municipal and industrial waste water. Local associations that are WPCF members are located in most states, and in five Canadian provinces. The Federation publishes books detailing all phases of waste water treatment, a technical journal, a newsletter, and public information pamphlets summarizing the disposal of sewage and hazardous wastes.

4. SUGGESTED READING

ASHWORTH, WILLIAM. *The Late, Great Lakes.* New York: Alfred A. Knopf, 1986.

The statistics applying to the Great Lakes are grand in scale: The Lakes, which hold one-fifth of the world's supply of fresh surface water, supply drinking water to twenty-four million people. Their coastline is *10,000 miles long*, with about half of it bordering the United States. Ashworth traces the ever-changing face of the Great Lakes—and the quality of their water—from their origin (in the last ice age) through the present.

ASHWORTH, WILLIAM. *Nor Any Drop to Drink*. New York: Summit Books, 1982.

"Nature persists in operating as a unit; humans persist in treating it as a collection of spare parts. Nature unifies, humans divide." This quote captures the spirit of Ashworth's book, which describes the basics of water supply and water pollution. He says the water crisis will persist until we learn to use the resource without overtaxing it.

BACH, JULIE S., AND HALL, LYNN. *The Environmental Crisis: Opposing Viewpoints*. St. Paul, MN: Greenhaven Press, 1986.

The book is part of a series of publications featuring opposing viewpoints on a variety of concerns of national importance. The series acts as a soapbox from which advocates for both sides of environmental issues can express their views, with the goal of finding solutions to problems of mutual concern. The views of environmentalists, health experts, government regulators, and the manufacturing industry are represented. The regulation of air and water pollution, the disposal of hazardous wastes, and acid rain are among the issues addressed in this volume.

BROWN, MICHAEL. *Laying Waste*. New York: Washington Square Press, 1981.

Working as a reporter during the late 1970s for the *Niagara Gazette*, Michael Brown wrote the first news stories about the apparent threat to health posed by Hooker Chemical's abandoned waste dump in the Love Canal neighborhood. In *Laying Waste*, Brown reviews the terrifying facts surrounding the discovery of Hooker's waste site at Love Canal and the subsequent discovery of even larger sites that were causing water contamination in the

Niagara Falls area. He goes on to describe similar instances of contamination nationwide.

EPSTEIN, SAMUEL S. *Hazardous Waste in America.* San Francisco: Sierra Club Books, 1982.

The kinds of hazardous waste to be found in this country and the ways in which they end up polluting the environment are outlined in the book. Federal laws governing toxic pollution are also explained. A "citizen's legal guide to hazardous wastes" explains how an individual can proceed legally to stop the violation of laws relating to the disposal of toxics.

FREUDENBERG, NICHOLAS. *Not in Our Backyards!: Community Action for Health and the Environment.* New York: Monthly Review Press, 1984.

The threat to health posed by the pollution of air and water by toxics, and the actions individuals and communities have taken to fight such contamination, are the focus of Freudenberg's book. Anyone who is already part of a group working to protect neighborhood water quality (or is interested in forming such a group) can find inspiration in the book by reading of the successful efforts of others and tips for getting started.

KAHAN, ARCHIE M. *Acid Rain: Reign of Controversy.* Golden, CO: Fulcrum, Inc., 1986.

The science and politics of acid rain are examined. Kahan presents the views of environmentalists, the coal and electric power industries, the scientific community, and government agencies to emphasize the controversy that surrounds the subject. The basics of how rainwater becomes acidic and acid rain's effects on plants and water creatures are explained.

KAHRL, WILLIAM L. *Water and Power.* Berkeley: University of California Press, 1982.

Kahrl's detailed history of the "big water grab"—the completion of Los Angeles Department of Water and Power's pipeline to the Owens Valley during the 1930s—is easy to read and full of life. The city flexed its considerable political muscle and employed cloak-

and-dagger tactics to secure the rights to most of the valley's water. It has been said that water flows toward money and power, and the Owens Valley story is a great illustration of that maxim.

KING, JONATHAN. *Troubled Waters*. Emmaus, PA: Rodale Press, 1985.

The story of the poisoning of America's drinking water—how it happened and its tragic effects—is effectively told in the book. Water polluted by toxic wastes and pesticides, deadly contaminants silently seeping from military installations, regulatory agencies too embroiled in politics to act to protect public health—these and other frightening dimensions of today's water crisis are vividly presented.

PALMER, TIM. *Endangered Rivers and the Conservation Movement*. Berkeley and Los Angeles: University of California Press, 1986.

The history of the harnessing of the nation's rivers to supply power and water for irrigation and drinking is outlined. The movement to clean up the nation's rivers and to preserve some of them in their natural state that emerged during the late 1960s and the 1970s is emphasized. Palmer closes with a description of the controversies that today surround the damming of rivers.

PLATT, RUTHERFORD. *Water: the Wonder of Life*. Englewood Cliffs, New Jersey: Prentice-Hall, 1978.

Human blood mimics the sea—where life originated. The blood in our bodies supplies nutrients to and carries wastes away from individual cells like the sea water in which unicellular life forms first developed. Platt's exploration of water and its many mysteries is engaging.

POWLEDGE, FRED. *Water*. New York: Farrar, Straus and Giroux, 1982.

Pork-barrel politics and exploitation have been key ingredients in the development of water resources in this country, according to Powledge, and as a result, we are faced with a growing shortage of potable water. Some of the "pork," some of the pollution, and some of the solutions to the crisis are described in his book.

PYE, VERONICA I. *Groundwater Contamination in the United States*. Philadelphia: University of Pennsylvania Press, 1983.

Groundwater, its movement, use, and pollution are described in the book. State, local, and federal regulations relating to the resource are summarized, and strategies for cleaning up aquifers and protecting them from pollution are also discussed.

REISNER, MARC. *Cadillac Desert*. New York: Viking, 1986.

The irrigation of arid regions, its cost, and its future are the themes of Reisner's easy-to-read book. The "Cadillac" in question is the network of dams, canals, and pipelines built at a cost of tens of billions of dollars of public funds to deliver water to these formerly dry lands. Reisner contends that much of the acreage that has been brought into production through federal irrigation projects will not produce enough in the long run to warrant the expense of delivering water to it. He says that drainage and salinity problems caused by heavy watering of desert soils are the bane of irrigated agriculture in arid climates.

SAFE DRINKING WATER COMMITTEE. *Drinking Water and Health*. Washington, D.C.: National Academy Press, vol. 1, 1977; vol. 2 & 3, 1980; vol. 4, 1982; vol. 5, 1983.

The Safe Drinking Water Act called for the National Academy of Sciences to initiate a series of studies on the health effects of toxic substances found in drinking water. The Safe Drinking Water Committee is a multi-disciplinary group of scientists and health professionals established by the Academy to summarize the results of that research.

Volume 1 contains estimates of the health risks associated with industrial chemicals, radioactive materials, and larger toxic elements and microorganisms that are carried by drinking water. The relative merits (and problems) associated with chlorination and other methods of water disinfection are compared, and the use of granular activated carbon filters is evaluated in volume 2. In volume 3, the health effects of trihalomethanes and other toxics formed in the water treatment process are estimated. Volume 4 takes a look at the chemical and biological contaminants associated with water-distribution systems. The health effects of the pesticides aldicarb and dinoseb and that of other toxics including arsenic, asbestos, carbon tetrachloride, trichloroethylene, uranium, and vinyl chloride are assessed in volume 5.

SCHEAFFER, JOHN R., AND STEVENS, LEONARD A. *Future Water.* New York: William Morrow and Co., 1983.

Waste water treatment systems that rely on the kind of natural, circular processes described in chapter 10 are featured in the book. John Scheaffer is a partner in a Chicago firm that has designed several such systems (including the one at the Hamilton Lakes development described in chapter 10).

SKJEI, ERIC, AND WHORTON, M. DONALD. *Of Mice and Molecules.* New York: The Dial Press, 1983.

What dangers does modern technology pose to our health? The authors take a look at the dark side of the technological revolution that has produced our modern world: the contamination of water and air with toxics that have an impact on health.

TOURBIER, J. TOBY, AND WESTMACOTT, RICHARD. *Resources Protection Technology.* Washington, D.C.: Urban Land Institute, 1981.

This "handbook of measures to protect water resources in land development" explains how urban developments can be designed to minimize their impact on water supply and quality. Measures to control runoff from paved areas, soil erosion and resulting sedimentation of streams, and pollution caused by heavy runoff from storms are among those discussed. Alternatives to conventional sewage treatment plants are discussed.

Resources Section B
Ranking of States by Volume of Hazardous Waste Generation and Hazardous Waste Program Quality

*T*he following table summarizes two 1980 studies:

- *Waste Generation Ranking*—These are the results of an EPA study that ranks states according to the total volume of hazardous wastes produced within their boundaries. The rankings will give you an idea of the overall level of industrial development within your state.
- *Program Quality Ranking*—The web of laws and officials whose job it is to protect a state's water resources is at least as important to water quality as is the total *volume* of wastes produced. State laws protecting water quality, the vigor with which they have been enforced, and the budget allotted to state water quality agencies were among the factors considered by the National Wildlife Federation in its ranking of the states' efforts to protect water quality. It should be noted that since the study was completed, some states, most notably Arizona (see chapter 11 for a description of Arizona's groundwater law) have made significant strides toward effective management of their water resources. The NWF study included the fifty states, the District of Columbia, and three U.S. territories.

253

HAZARDOUS WASTE RANKINGS		
State	Waste Generation	Program Quality
New Jersey	1	12
Ohio	2	10
Illinois	3	17
California	4	1
Pennsylvania	5	15
Texas	6	13
New York	7	16
Michigan	8	28
Tennessee	9	3
Indiana	10	25
North Carolina	11	14
Virginia	12	23
Missouri	13	39
Louisiana	14	18
South Carolina	15	4
Massachusetts	16	31
Florida	17	9
Wisconsin	18	30
West Virginia	19	43
Georgia	20	48
Connecticut	21	8
Kentucky	22	42
Alabama	23	19
Maryland	24	2
Minnesota	25	32
Washington	26	5
Iowa	27	35
Kansas	28	20
Delaware	29	6
Mississippi	30	45
Arkansas	31	22
Colorado	32	41
Oklahoma	33	21
Oregon	34	7

State	Waste Generation	Program Quality
Rhode Island	35	26
Idaho	36	46
Maine	37	44
Nebraska	38	40
Arizona	39	53
New Hampshire	40	49
Utah	41	36
New Mexico	42	33
Montana	43	29
Vermont	44	11
Nevada	45	51
Alaska	46	37
Washington, D.C.	47	27
Hawaii	48	38
North Dakota	49	34
South Dakota	50	52
Wyoming	51	24

SOURCES: Program-quality ranking: EPA grant #900905-01 (1980), National Wildlife Federation; Waste generation ranking, EPA study (1980).

With the large number of instances of toxic waste disposal facilities polluting local water sources, it's good to know whether there's a major waste-handling station in your area. The possibility of spills from trucks hauling waste to the disposal center or leaks from the facility itself increase with your proximity to the site. Increasingly stringent regulation of waste disposal operations in the last few years has caused many of the worst polluters to close rather than meet the tougher federal regulations, but that doesn't mean the remaining facilities aren't a potential threat to water quality. The following list is from a 1987 EPA report on the disposal of the nation's hazardous wastes.

OPERATING COMMERCIAL INCINERATOR FACILITIES*			
Owner	**Location**	**Type of Unit**	**Type of Wastes**
Environmental Systems Company	El Dorado Arkansas	Rotary Kiln	PCB, Acids, Halogenated & Non-Halogenated Solvents, Halogenated & Non-Halogenated Organics
International Technology Corporation	Martinez California	Liquid Injection	Acids, Non-Halogenated Solvents & Organics, Metallic Inorganics
Chemical Waste Management Inc.	Sauget Illinois	Liquid Injection & Fixed Hearth	Halogenated & Non-Halogenated Solvents, Halogenated & Non-Halogenated Organics
Chemical Services, Inc.	Chicago Illinois	Liquid Injection & Rotary Kiln	PCB, Halogenated & Non-Halogenated Solvents, Halogenated & Non-Halogenated Organics, Non-Metallic Inorganics
LWD, Inc.	Calvert City Kentucky	Liquid Injection	Acids, Halogenated & Non-Halogenated Solvents, Halogenated & Non-Halogenated Organics, Metallic Organics
LWD, Inc.	Clay Kentucky	Rotary Kiln	Acids, Halogenated & Non-Halogenated Solvents, Halogenated & Non-Halogenated Organics, Metallic Organics

*In addition, there are four TSCA commercial incinerators permitted to burn PCB wastes. They include: Pyrochem (Coffeyville, Kansas), Pyrotech Systems—mobile unit, U.S. EPA incinerator—mobile unit, and General Electric (Pittsfield, Massachusetts).

SOURCE: *The Hazardous Waste System,* EPA Office of Solid Waste and Emergency Response (June, 1987).

Owner	Location	Type of Unit	Type of Wastes
Rollins Environmental Services	Baton Rouge Louisiana	Liquid Injection & Rotary Kiln	Acids, Halogenated & Non-Halogenated Solvents, Halogenated & Non-Halogenated Organics, Metallic Organics, Metallic and Non-Metallic Inorganics
Rollins Environmental Services	Bridgeport New Jersey	Liquid Injection & Rotary Kiln	Acids, Halogenated & Non-Halogenated Solvents, Halogenated & Non-Halogenated Organics, Metallic Organics, Metallic and Non-Metallic Inorganics
Rollins Environmental Services	Deer Park Texas	Liquid Incineration & Rotary Kiln	PCB, Acids, Halogenated & Non-Halogenated Solvents, Halogenated & Non-Halogenated Organics, Metallic Organics, Metallic and Non-Metallic Inorganics
Caldwell Systems, Inc.	Lenoir North Carolina	Liquid Injection & Solid Incineration	Halogenated & Non-Halogenated Solvents, Halogenated & Non-Halogenated Organics, Metallic & Non-Metallic Organics
Ross Incineration	Grafton Ohio	Liquid Injection & Rotary Kiln	Acids, Halogenated & Non-Halogenated Solvents, Halogenated & Non-Halogenated Organics
Stablex South Carolina Inc.	Rock Hill South Carolina	Fixed Hearth	Halogenated & Non-Halogenated Solvents, Halogenated & Non-Halogenated Organics, Metallic Organics
GSX Thermal Oxidation Corp.	Roebuck South Carolina	Liquid Injection	Halogenated & Non-Halogenated Solvents, Halogenated & Non-Halogenated Organics
B.D.T., Inc.	New York	Not Available	Metals

OPERATING COMMERCIAL LAND DISPOSAL FACILITIES

Owner	Location	Type of Facilities	Wastes
Chemical Waste Management Inc	Emelle Alabama	Landfill, Storage Impoundments, Treatment Impoundments	Metals, Cyanides, Acidic Corrosives, PCBs, Halogens
Lion Oil Company	El Dorado Arizona	Land Treatment, Storage Impoundments	Metals
IT Corp Benecia	Benecia California	Landfill, Disposal Impoundments, Storage Impoundments	Metals, Cyanides, Solvents
IT Corp Vine Hill	Martinez California	Treatment Impoundments	Metals
IT Corp Imperial	Westmoreland California	Disposal Impoundments, Treatment Impoundments	Metals, Solvents
Casmalia Resources	Casmalia California	Landfill, Disposal Impoundments, Treatment Impoundments	Acidic Corrosives, Metals, Cyanides, Halogens
Chemwest Industries Inc	Fontana California	Storage Impoundments	Acidic Corrosives
AMCE Fill Corporation	Martinez California	Landfill	Other
IT Corp Baker Facility	Martinez California	Disposal Impoundments, Treatment Impoundments	Metals, Acidic Corrosives
Chemical Waste Management Inc	Kettleman City California	Landfill, Treatment Impoundments	Acidic Corrosives, Metals
CECOS International Inc	Bristol Connecticut	Waste Piles	Metals, Cyanides
City of Danbury	Danbury Connecticut	Landfill	Metals
Torrington Landfill	Torrington Connecticut	Landfill	Metals

Owner	Location	Type of Facilities	Wastes
Salsbury Laboratories	Charles City Iowa	Storage Impoundments, Treatment Impoundments	Metals, Solvents, Halogens
Envirosafe Services of Idaho	Grand View Idaho	Landfill, Waste Piles	Acidic Corrosives, Metals, Cyanides, Solvents, PCBs, Halogens
SCA Chemical Services Inc	Chicago Illinois	Storage Impoundments, Treatment Impoundments	Other
Peoria Disposal Co	Peoria Illinois	Landfill	Metals
CID-Landfill	Calumet City Illinois	Landfill	Acidic Corrosives, Metals, Cyanides, Solvents, Halogens
Kerr-McGee Chemical Corp	Madison Illinois	Storage Impoundments	Other
CECOS International Inc./ BFI	Zion Illinois	Landfill	Metals, Solvents, Halogens
Four County Landfill	Rochester Indiana	Landfill	Metals
Adams Center Landfill Inc	Fort Wayne Indiana	Landfill	Acidic Corrosives, Metals, Cyanides, Solvents, Halogens
CECOS International Inc.	Westlake Louisiana	Storage Impoundments	Acidic Corrosives, Metals, Solvents, Halogens
CECOS International Inc.	Livingston Louisiana	Landfill	Acidic Corrosives, Cyanides, Solvents, Halogens
Chemical Waste Management Inc	Carlyss Louisiana	Landfill	Metals, Cyanides, Solvents, Halogens

OPERATING COMMERCIAL LAND DISPOSAL FACILITIES (cont.)			
Owner	**Location**	**Type of Facilities**	**Wastes**
Rollins Environmental Services	Baton Rouge Louisiana	Landfill, Storage Impoundments, Treatment Impoundments	Metals, Solvents, Cyanides, Acidic Corrosives
Wayne Disposal, Inc	Bellerville Michigan	Treatment Impoundments	Acidic Corrosives, Metals
Environmental Waste Control	Inkster Michigan	Treatment Impoundments	Acidic Corrosives, Metals
Chem-Met Services Inc	Wyandotte Michigan	Waste Piles	Acidic Corrosives, Metals, Solvents, Halogens
Federal-Hoffman Inc	Anokia Minnesota	Landfill	Other
North Star Steel Co	St. Paul Minnesota	Waste Piles	Metals
B. H. S. Inc	Wright City Missouri	Landfill	Metals, Halogens
Rogers Rental Landfill	Centreville Mississippi	Land Treatment	Other
Burlington Northern Somers	Somers Montana	Waste Piles, Storage Impoundments	Acidic Corrosives, Metals, Solvents
US Ecology Inc	Beatty Nevada	Landfill	Metals, Cyanides, Solvents, PCBs, Halogens
Frontier Chemical Waste Process	Niagara Falls New York	Waste Piles	Metals
CECOS International Inc	Niagara Falls New York	Landfill	Acidic Corrosives, Metals, PCBs
F E I Landfarming	Oregon Ohio	Land Treatment	Metals
Ashland Chemical Co	South Point Ohio	Waste Piles	Other

Owner	Location	Type of Facilities	Wastes
Chemical Waste Management Inc	Vickery Ohio	Storage Impoundments	Acidic Corrosives, Metals
Fondessy Enterprises Inc	Oregon Ohio	Landfill	Metals, Cyanides, Solvents, Halogens
Erieway Pollution Control Inc	Bedford Ohio	Waste Piles	Acidic Corrosives, Metals, Halogens
CECOS International Inc	Williamsburg Ohio	Landfill	Metals, Cyanides, Solvents, PCBs, Halogens
Delhi Industrial Products	McDonald Ohio	Waste Piles	Metals
Eagle Picher Industries Inc	Quapaw Oklahoma	Disposal Impoundments	Metals, Solvents
USPCI	Waynoka Oklahoma	Landfill, Disposal Impoundment, Waste Piles, Storage Impoundments, Treatment Impoundments	Acidic Corrosives, Metals, Cyanides
Chem-Security Systems Inc	Arlington Oregon	Landfill, Storage Impoundments, Treatment Impoundments	Acidic Corrosives, Metals, Solvents, PCBs, Halogens
Mill Service Inc	Yukon Pennsylvania	Disposal Impoundments, Waste Piles	Metals
Mill Service Inc	Bulger Pennsylvania	Disposal Impoundments, Waste Piles	Metals
GSX Services of South Carolina	Pinewood South Carolina	Storage Impoundments	Acidic Corrosives, Metals, Cyanides
Yale Security Inc	Lenoir City Tennessee	Storage Impoundments	Acidic Corrosives, Metals
Gibraltar Chemical Resources	Winona Texas	Storage Impoundments	Acidic Corrosives, Metals

OPERATING COMMERCIAL LAND DISPOSAL FACILITIES (cont.)

Owner	Location	Type of Facilities	Wastes
Gulf Coast Waste Disposal	Texas City Texas	Landfill, Land Treatment	Metals, Cyanides
Chemical Waste Management Inc	Port Arthur Texas	Landfill, Disposal Impoundments, Storage Impoundments, Treatment Impoundments	Acidic Corrosives, Metals, Cyanides, Solvents, Dioxins, Halogens
Rollins Environmental Services	Deer Park Texas	Landfill, Storage Impoundments, Treatment Impoundments	Metals, Cyanides, Solvents, Halogens
Olin Corporation	Beaumont Texas	Treatment Impoundments	Acidic Corrosives, Metals, Cyanides, Solvents, Halogens
Malone Service Company	Texas City Texas	Landfill, Storage Impoundments, Treatment Impoundments	Metals, Cyanides, Acidic Corrosives
Texas Ecologists Inc.	Robstown Texas	Landfill	Metals, Cyanides, Solvents, Halogens
USPCI	Knowles Utah	Landfill, Land Treatment, Storage Impoundment	Metals, Acidic Corrosives, Solvents, PCBs, Halogens

SOURCE: EPA Office of Solid Waste

Resources Section C
State Water Quality and Health Agencies

ALABAMA

Department of Environmental
Management
1751 Federal Drive
Montgomery, AL 36130
(205) 271-7700

Department of Public Health
381 State Office Bldg.
Montgomery, AL 36130
(205) 261-5052

ALASKA

Department of Natural
Resources, Div. of Land
and Water
Pouch M
Juneau, AK 99811
(907) 465-2400

Department of Health and
Social Services, Div. of
Public Health
Alaska Office Bldg., Pouch
H-01
Juneau, AK 99811
(907) 465-3030

ARIZONA

Department of Water
Resources
99 E. Virginia Ave.
Phoenix, AZ 85004
(602) 255-1554

Department of Health
Services
1740 West Adams St.
Phoenix, AZ 85007
(602) 255-1024

ARKANSAS

Department of Commerce
1 Capital Mall, Suite 2D
Little Rock, AR 72201
(501) 371-1611

Department of Health
4815 W. Markham St.
Little Rock, AR 72201
(501) 661-2111

CALIFORNIA

Department of Water
Resources
PO Box 388
Sacramento, CA 95802
(916) 455-9248

Department of Health
Services
714 P St.
Sacramento, CA 95814
(916) 445-1102

COLORADO

Department of Natural
Resources, Water
Resources Division
4210 E. 11th Ave.
Denver, CO 80220
(303) 866-3587

Department of Health
4210 E. 11th Ave.
Denver, CO 80220
(303) 320-8333

CONNECTICUT

Department of Environmental
Protection
165 Capital Ave., Rm 553
Hartford, CT 06106
(203) 566-3540

Department of Health
150 Washington St.
Hartford, CT 06115
(203) 566-2279

DELAWARE

Department of Natural
Resources and
Environmental Control
PO Box 1401
Dover, DE 19903
(302) 736-4403

Department of Health and
Social Services
Jesse S. Cooper Memorial
Bldg.
Dover, DE 19901
(302) 678-4731

FLORIDA

Department of Environmental
Regulation
2600 Blair Stone Rd.
Tallahassee, FL 32301
(904) 488-4805

Department of Health and
    Rehabilitative Services
1317 Winewood Blvd.
Tallahassee, FL 32301
(904) 488-7721

GEORGIA

Department of Natural
    Resources
270 Washington St., SW
Atlanta, GA 30334
(404) 656-3500

Division of Physical Health
47 Trinity Ave., SW
Atlanta, GA 30334
(404) 656-4734

HAWAII

Department of Land and
    Natural Resources
PO Box 621
Honolulu, HI 96809
(808) 548-6550

Department of Health
1250 Punchbowl St.
Honolulu, HI 96801
(808) 548-6505

IDAHO

Department of Water
    Resources
State House
Boise, ID 83720
(208) 334-2190

Department of Health and
    Welfare
450 W. State St.
Boise, ID 83720
(208) 334-4079

ILLINOIS

Illinois Environmental
    Protection Agency
2200 Churchill Rd.
Springfield, IL 62706
(217) 782-1654

Department of Public Health
535 W. Jefferson St.
Springfield, IL 62761
(217) 782-4997

INDIANA

Department of Natural
    Resources
608 State Office Bldg.
Indianapolis, IN 46204
(317) 232-4020

Board of Health
1330 W. Michigan St.
Indianapolis, IN 46206
(317) 633-8400

IOWA

Department of Water, Air and
    Waste Management
900 E. Grand Ave.
Des Moines, IA 50319
(515) 281-8690

Department of Health
E. 12th and Walnut Sts.
Des Moines, IA 50319
(515) 281-5605

KANSAS

Department of Health and
  Environment
Forbes Field
Topeka, KS 66620
(913) 862-9360
for both water quality and
  health

KENTUCKY

Department for Natural
  Resources and
  Environmental Protection
18 Reilly Rd.
Frankfort, KY 40601
(502) 564-2150

Department for Human
  Resources
275 E. Main St.
Frankfort, KY 40621
(502) 564-3970

LOUISIANA

Department of Natural
  Resources
PO Box 44066
Baton Rouge, LA 70804
(504) 342-1266

Department of Health and
  Human Services
PO Box 60630
New Orleans, LA 70160
(504) 568-5100

MAINE

Department of Environmental
  Protection
State House, Stn. 17
Augusta, ME 04333
(207) 289-2811

Department of Human
  Services, Bureau of Health
State House, Stn. 11
Augusta, ME 04333
(207) 289-3201

MARYLAND

Department of Natural
  Resources
580 Taylor Ave.
Annapolis, MD 21401
(301) 269-3041

Department of Health and
  Mental Hygiene
201 W. Preston St., 5th Floor
Baltimore, MD 21201
(301) 383-6195

MASSACHUSETTS

Office of Environmental
  Affairs
100 Cambridge St.
Boston, MA 02202
(617) 727-9800

Office of Human Services
600 Washington St.
Boston, MA 02111
(617) 727-2700

MICHIGAN

Department of Natural
  Resources
PO Box 30028
Lansing, MI 48909
(517) 373-2329

Department of Public Health
3500 N. Logan St.
Lansing, MI 48909
(517) 373-1320

MINNESOTA

Department of Natural
  Resources
658 Cedar St.
St. Paul, MN 55146
(612) 296-2549

Department of Health
717 Delaware St., SE
Minneapolis, MN 55440
(612) 623-5000

MISSISSIPPI

Department of Natural
  Resources
PO Box 20305
Jackson, MS 39209
(601) 961-5000

Board of Health, Bureau of
  Environmental Health
PO Box 1700
Jackson, MS 39205
(601) 354-6646

MISSOURI

Department of Natural
  Resources
1915 Southridge Plaza
Jefferson City, MO 65102
(314) 751-4422

Department of Social Services
221 W. Itish St.
Jefferson City, MO 65102
(314) 751-4330

MONTANA

Department of Health &
  Environmental Science,
  Water Quality Bureau
32 S. Ewing
Helena, MT 59620
(406) 444-2544
for both health and water
  quality

NEBRASKA

Department of Environmental
  Control
PO Box 94877
(402) 471-2186

Department of Health
301 Centennial Mall, S., 3rd
  Floor
Lincoln, NE 68509
(402) 471-2133

NEVADA

Department of Conservation
 and Natural Resources
201 S. Fall St.
Carson City, NV 89710
(702) 885-4670

Department of Human
 Services
505 E. King St.
Carson City, NV 89710
(702) 885-4740

NEW HAMPSHIRE

Office of State Planning, Div.
 of Water Supply
2½ Beacon St.
Concord, NH 03301
(603) 271-1110

Department of Health and
 Welfare
Hazen Dr.
PO Box 95
Concord, NH 03301
(603) 271-4501

NEW JERSEY

Department of Environmental
 Protection
Labor and Industry Building,
 Rm. 802
Trenton, NJ 08625
(609) 292-2885
for both health and water
 quality

NEW MEXICO

Department of Health and
 Environment
PO Box 968
Santa Fe, NM 87503
(505) 948-0020
for both health and water
 quality

NEW YORK

Department of Environmental
 Conservation
50 Wolf Rd.
Albany, NY 12233
(518) 457-6934

Department of Health
Tower Building, Room 482
Albany, NY 12207
(518) 474-2011

NORTH CAROLINA

Department of Natural
 Resources
512 N. Salisbury St.
Raleigh, NC 27611
(919) 733-4984

Department of Human
 Resources
225 N. McDowell St.
Raleigh, NC 27602
(919) 733-3446

NORTH DAKOTA

State Water Commission
900 East Blvd.
Bismark, ND 58505
(701) 224-2750

Department of Health
State Capitol
Bismark, ND 58505
(701) 224-2180

OHIO

Department of Natural
    Resources
Fountain Square, Bldg. E3
Columbus, OH 43224
(614) 265-6877

Department of Health
246 N. High St.
Columbus, OH 43215
(614) 466-2253

OKLAHOMA

Department of Health, Water
    Quality Service
100 NE 10th
Oklahoma City, OK 73152
(405) 271-4200
both health and water quality

OREGON

Water Resources Department
555 13th St. NE
Salem, OR 97310
(503) 378-3739

State Health Division, Public
    Health Laboratory
1717 SW 10th Ave.
Portland, OR 97201
(503) 229-5882

PENNSYLVANIA

Department of Environmental
    Resources
PO Box 2063
Harrisburg, PA 17108
(717) 787-2814

Department of Health
PO Box 90
Harrisburg, PA 17108
(717) 783-3795

RHODE ISLAND

Department of Environmental
    Management
38 Park St.
Providence, RI 02903
(401) 277-2771

Department of Health
75 Davis St.
Providence, RI 02908
(401) 277-2231

SOUTH CAROLINA

Water Resources Commission
3830 Forest Dr.
Columbia, SC 29240
(803) 758-2514

Department of Health and
    Environmental Control
2600 Bull St.
Columbia, SC 29201
(803) 758-5443

SOUTH DAKOTA

Department of Water and
Natural Resources
523 E. Capital Ave.
Pierre, SD 57501
(605) 773-3151
for both health and water
quality

TENNESSEE

Department of Health and
Environment
150 9th Ave., N
Nashville, TN 37219
(615) 741-3111
both health and water quality

TEXAS

Department of Water
Resources
Box 13087, Capitol Station
Austin, TX 78711
(512) 475-7036

Department of Health, Div. of
Water Hygiene
1100 W. 49th St.
Austin, TX 78756
(512) 458-7542

UTAH

Department of Natural
Resources and Energy
1636 W. North Temple
Salt Lake City, UT 84116
(801) 533-5356

Department of Health
150 W. North Temple
Salt Lake City, UT 84113
(801) 533-6111

VERMONT

Environmental Conservation
Agency
79 River St.
Montpelier, VT 05602
(802) 828-3130

Department of Health
60 Main St.
Burlington, VT 05402
(802) 863-7200

VIRGINIA

State Water Control Board
2111 Hamilton St.
Richmond, VA 23230
(804) 257-0056

Department of Human
Resources
109 Governor St.
Richmond, VA 23219
(804) 786-3561

WASHINGTON

Department of Ecology, Office
of Water Programs
Mailstop PV-11
Olympia, WA 98504
(206) 459-6168

Department of Social and
Health Services
Mailstop LD-11
Olympia, WA 98504
(206) 753-3395

Department of Health and
Social Services
Box 7850
Madison, WI 53707
(608) 266-3681

WEST VIRGINIA

Department of Natural
Resources
1800 Washington St., E.
Charleston, WV 25305
(304) 348-2754

Department of Health
1800 Washington St., E.
Charleston, WV 25305
(304) 348-2971

WISCONSIN

Department of Natural
Resources
PO Box 7921
Madison, WI 53707
(608) 266-2121

WYOMING

Department of Environmental
Quality
Herschler Bldg.
Cheyenne, WY 82002
(307) 777-7937

Division of Health and
Medical Services
Hathaway Bldg.
Cheyenne, WY 82002
(307) 777-7431

Resources Section D
Superfund Toxic Waste Sites

*H*azardous waste sites on the Superfund National Priorities List (NPL) are a *known* threat to public health. There are currently 802 sites on the list, and the EPA has proposed another 149 be added. A total of more than 26,000 sites is being reviewed for possible inclusion on the NPL.

The following sites are either on the list of existing or proposed NPL sites. Those reviewed in this volume were selected either because they represent a severe threat to drinking water quality, or because they are representative of a class of pollution sources. As pointed out in the text, the number of Superfund sites in a state doesn't necessarily reflect the seriousness of water contamination within the state's boundaries, since more will be known about potential pollution sources in states with more aggressive water quality protection programs.

SUPERFUND TOXIC WASTE CLEANUP PROGRAM
National Priorities List,
Final and Proposed Sites Per State/Territory
(by Total Sites)
July 1987

	Total
New Jersey	100
Pennsylvania	80
Michigan	69
New York	67
California	63
Minnesota	40
Florida	39
Wisconsin	33
Ohio	30
Indiana	29
Washington	28
Illinois	27
Texas	26
Massachusetts	21
Virginia	19
Missouri	19
Delaware	18
South Carolina	15
Colorado	15
New Hampshire	13
Iowa	13
North Carolina	11
Utah	10
Tennessee	10
Kentucky	10
Arkansas	10
Alabama	10
Montana	9
Maryland	9
Arizona	9
Rhode Island	8

	Total
Puerto Rico	8
Louisiana	8
Kansas	8
Connecticut	8
Maine	7
Georgia	7
West Virginia	6
Oregon	6
Oklahoma	6
Hawaii	6
Nebraska	5
New Mexico	4
Idaho	4
Vermont	2
Mississippi	2
Wyoming	1
South Dakota	1
North Dakota	1
Guam	1
Virgin Islands	0
Trust Territories	0
Nevada	0
District of Columbia	0
Commonwealth of Marianas	0
American Samoa	0

SOURCE: EPA.

ALABAMA

Alabama Army Ammunition Plant
Childersberg, Alabama
Explosives were manufactured until 1945 on the 5,100-acre site, just east of the Coosa River north of Childersberg, in Talladega County. Spills of some of the toxic elements used in the process and

several waste dumps on the site are the primary causes of contamination of nearby surface and groundwater; both are used locally as drinking water. The Department of Defense and the Army are studying ways in which the site can be cleaned up.

Anniston Army Depot
Anniston, Alabama
The thirty-square-mile depot has been used to store ammunition since 1941. Combat vehicles and artillery equipment are also repaired at the base. Metals and solvents disposed of in five lagoons on the base have contaminated the underlying aquifer. The Anniston municipal water supply, which furnishes water to 39,000 people, may be threatened with contamination as a result. A study of the extent of contamination and potential remedial action is under way.

Interstate Lead Company
Leeds, Alabama
The company operates a lead-smelting and battery recycling operation that has produced tens of thousands of tons of hazardous wastes over the years. Some of the lead-bearing wastes are located on the company's property, and some have been used as fill under some of the buildings in the city. Soil and groundwater have become contaminated as a result. Some of the wastes have been removed and some have been covered with a synthetic liner to slow the spread of contamination. In 1988 Interstate Lead was going through bankruptcy proceedings.

Triana/Tennessee River
Limestone & Morgan Counties, Alabama
Residents of Triana have been found by the U.S. Centers for Disease Control to have higher-than-average concentrations of DDT in their bodies. Olin Corporation manufactured DDT (under an Army contract) at the Redstone Arsenal on the Tennessee River in Huntsville from 1947 to 1970. Wastes from the operation were routed to the river and an estimated 475 tons of DDT residues are now part of its sediments, as a result. The Olin Corporation was required in the 1983 settlement of a federal lawsuit to take steps to minimize human exposure to the DDT still in the river.

ARIZONA

Motorola, Incorporated (52nd St. Plant)
Phoenix, Arizona
Semiconductors are manufactured at the plant, which is a mile
and a half northeast of Sky Harbor International Airport. The
solvents trichloroethylene (TCE) and trichloroethane (TCA) have
contaminated groundwater under the airport and at least a mile to
the west. The site has been proposed for addition to the NPL.
Further study of the extent of pollution is being carried out.

Tucson International Airport
Tucson, Arizona
The site covers about twenty-four square miles in the vicinity of
the airport in southwest Tucson. The airport has been a center for
the construction and modification of military aircraft, missiles, and
sophisticated electronics systems since the early 1950s. The solvent TCE, hexavalent chromium, chloroform, and a variety of other
toxic industrial chemicals have polluted groundwater in a four-
square-mile area (reaching northeast from the airport, parallel to
the Santa Cruz River) as the result of spills, leaks, and the use of
degreasers. Groundwater is the only source of drinking water for
the 600,000 people in the Tucson metropolitan area. Eight city
wells have been shut down as a result of the contamination. The Air
Force has agreed to spend $28 million to clean up a portion of the
contamination, and is paying the city to install aeration equipment
on the polluted wells so they can be reopened.

ARKANSAS

Jacksonville and Rogers Road Municipal Landfills
Jacksonville, Arkansas
The Jacksonville Municipal Landfill is an eighty-acre dump on
Graham Road just inside Lonoke County, which operated from 1960
until 1973. A variety of extremely toxic wastes including dioxin,
DDT, PCBs, and heptachlor epoxide were disposed of there (in
drums and in unlined seepage pits) until the dump was closed

down by the state. The dump is half a mile from the Rogers Road Municipal Landfill, also a proposed addition to the National Priority List. Dioxin, TCE, 2,4-D, 2,4,5-T, and PCBs were dumped at this site from 1953 until it was closed in 1974. The EPA is planning an investigation into the extent of the contamination caused by both dumps. Wells within three miles supply drinking water to an estimated 10,000 people, and the shallow aquifer in the area has been found to have been polluted.

CALIFORNIA

THE SILICON VALLEY

In the "Silicon Valley," located south of California's San Francisco Bay, a total of nineteen NPL sites, primarily semiconductor manufacturing businesses, are polluting groundwater used by hundreds of thousands of people in this heavily developed area. A few sites are reviewed here, and some of the rest are more briefly listed. The California Department of Health Services and the California Regional Water Quality Control Board are overseeing an investigation of the extent of pollution and the cleanup.

Applied Materials
Santa Clara, California
Applied Materials makes chips for the electronics industry on a 2.5-acre site. Groundwater under the site is contaminated, primarily with industrial solvents that leaked out of underground storage tanks. Approximately 300,000 people use groundwater drawn from within three miles of the site. The company is pumping some of the contaminated water, treating it, and returning it to the aquifer.

Castle Air Force Base
Merced, California
The disposal of solvents, cyanide, cadmium, fuels, and waste oil in unlined pits and landfills on the 2,700-acre base have polluted groundwater. The Air Force (which had to drill a new, deeper well in 1984 when TCE was found in its 300-foot-deep well) is con-

ducting an investigation into the extent of contamination and ways of cleaning it up.

Moffett Naval Air Station
Sunnyvale, California
Solvents have been found in the aquifer underlying the 8,700-acre air base. An estimated 272,000 people use water from wells located within three miles. The extent of the contamination caused by the site is being studied.

Monolithic Memories, Incorporated
Sunnyvale, California
Leaking underground storage tanks have contaminated groundwater with chloroform, xylene, and the solvent TCE. An estimated 300,000 people drink groundwater within three miles of the company's twenty-acre site. The company is pumping groundwater from under the site and treating it while an investigation of the extent of contamination caused is under way.

Fairchild Camera and Instrument Corporation
Mountain View, California
Groundwater is polluted with TCE and other solvents.

Hewlett-Packard
Palo Alto, California
TCE, TCA, and toluene are among the contaminants found in groundwater under the plant.

IBM Corporation
San Jose, California
TCE, TCA, Freon 113, and other industrial solvents have polluted groundwater under the plant.

National Semiconductor Corporation
Sunnyvale, California
Groundwater contamination by vinyl chloride, TCE, and other solvents.

Teledyne Semiconductor
Mountain View, California
Groundwater is contaminated by solvents, including TCE.

OTHER CALIFORNIA NPL SITES

FMC Corporation
Fresno, California
Groundwater below the company's grounds has become contaminated with heavy metals and pesticides as the result of thirty years' production of pesticides on the site. Wells for the Fresno municipal water-supply system, which serves 250,000 people, are near the site. The site has been proposed for addition to the NPL. An investigation into the extent of contamination and possible cleanup alternatives is being conducted.

Lawrence Livermore National Laboratory
Livermore, California
The site was first used as a Naval Air Station during the 1940s; hazardous materials have long been stored, used, and disposed of on the one-square-mile site, located three miles west of town. Since 1952 it has been operated as a nuclear weapons and fusion energy research center, first by the Atomic Energy Commission and then by the Department of Energy. Groundwater in the area has been found to be contaminated with solvents, petroleum products, and a variety of chemicals coming from two on-site waste disposal pits, leaking underground storage tanks, and outdoor chemical storage areas on the grounds. LLNL has drilled 160 monitoring wells on and around the site, and ways of cleaning up the contamination, which has been detected more than half a mile away, are being studied.

Louisiana-Pacific Corporation
Oroville, California
PCP is sprayed on lumber from the mill as a preservative. Leaks and spills have resulted in pollution of both the shallow and a deeper aquifer underlying the site. An estimated 10,500 people use water from wells within three miles of the site. Further investigation into the extent of contamination and into potential ways of cleaning it up is being conducted.

McClellan Air Force Base
Sacramento, California
Toxic wastes dumped at thirty-six sites on the 2,600-acre base,

located eight miles northeast of Sacramento, have polluted the underlying aquifer. The contamination (caused primarily by solvents used in the maintenance and repair of airplanes) has forced the closure of eleven private wells and one municipal well (serving an estimated 23,000 people) since 1979. The Air Force is currently investigating the extent of the contamination and ways of cleaning it up; it recently paid to have 500 nearby residences that had relied on groundwater hooked up to the municipal water system.

Southern California Edison Company
Visalia, California
Utility poles were treated with pentachlorophenol (PCP) and creosote on the site from the 1920s to 1980. Spills, leaking storage tanks, and drips from treated poles have polluted groundwater under the site with these chemicals and with dioxin, which is a component of the PCP. The company has installed an underground slurry wall to slow the migration of contaminated water from the site, and is pumping and filtering polluted water from the shallow aquifer underlying the site. California Water Service Company wells within three miles of the site provide water to 59,000 people in Visalia.

Stringfellow
Glen Avon Heights, California
The Stringfellow acid pits are California's highest priority NPL site. Between 1956 and 1972 an estimated thirty-four million gallons of liquid industrial wastes were disposed of on the twenty-two-acre site, in a canyon near Glen Avon Heights. Acids, toxic organic chemicals, and heavy metals in the wastes have contaminated both surface and groundwater in the area. In 1980, EPA's Emergency Response Team removed approximately ten million gallons of wastes from the site. In 1981, the state reduced the flow of contaminants by capping the site and installing drainage controls.

Waste Disposal, Incorporated
Santa Fe Springs, California
The company operated a waste disposal operation in Santa Fe Springs (ten miles east of Los Angeles) from 1928 to 1965. A variety of industrial wastes were disposed of on the site. Benzene, benzo-a-

pyrene, phenol, and toluene have been found in soil under the site, and an investigation of possible groundwater contamination is under way.

COLORADO

Martin Marietta
Waterton, Colorado
The 5,200-acre Martin Marietta Denver Aerospace plant southeast of Denver has been in operation since 1956. Hexavalent chromium, solvents, and a variety of other toxic chemicals that were disposed of in waste lagoons on the property have been found in area groundwater. An intake that supplies about 15 percent of Denver's municipal water is a mile and a half down-gradient (in the direction of waterflow in the aquifer) from the disposal areas. More than one million customers use Denver water. Investigation of the extent of pollution and of ways of cleaning it up is now in progress.

Rocky Flats Plant
Golden, Colorado
Nuclear weapons components have been produced on the 6,550-acre site since 1951. Plutonium and tritium that were buried there have been found in ground and surface water, polluting the water supply of the city of Broomfield and others. Cleanup of hot spots and analysis of the extent of water contamination is being conducted by the U.S. Department of Energy, the owner of the facility.

Rocky Mountain Arsenal
Denver, Colorado
The Arsenal (described in chapter 4) is believed by many to be the site of some of the most massive and dangerous groundwater contamination in the country. The Army and Shell Oil have already spent more than $100 million attempting to deal with contamination issuing from the twenty-seven-square-mile Adams County site, but have hardly scratched the surface.

Uravan Uranium Mill
Uravan, Colorado
The mill, located on the San Miguel River five miles upstream

from its confluence with the Dolores River, started operation in 1915, processing radium. Uranium milling started during the 1940s, when the market for radium dried up. Uranium from the site was used in the nation's first atomic bombs, and later in nuclear power plants. The mine was closed in 1984. An estimated ten million tons of tailings from the mine have contaminated the San Miguel and Dolores Rivers and local groundwater. The owner, Union Carbide, agreed in 1987 to spend $40 million cleaning up the site.

CONNECTICUT

Revere Textile Prints Corporation
Sterling, Connecticut
A textile plant that operated on the fifty-acre site from the 1930s until it burned down in 1980 has caused contamination of groundwater with solvents and of the Moosup River with a variety of pollutants. Groundwater is the sole source of water for approximately 4,500 people in the area.

DELAWARE

Dover Gas and Light
Dover, Delaware
The utility operated a plant on the site between 1859 and 1948 that produced gas from coal to light the city's streetlamps. When natural gas became available, the plant was dismantled, and wastes including coal-tar residues were buried on the site. Geologic studies preceding the construction of a new county courthouse on the site in 1984 revealed that toxic chemicals from the wastes (including benzene, toluene, xylene, lead, and polynuclear aromatic hydrocarbons) were entering groundwater. Seven of Dover's fourteen municipal wells are within a mile of the site. Further studies of the extent and sources of contamination are being conducted.

E.I. du Pont de Nemours and Company, Incorporated
Newport, Delaware
The company operated a seven-acre landfill next to its paint-
pigment plant from 1902 to 1975. Heavy metals including barium,
cadmium, and zinc, as well as solvents, have contaminated aquifers underlying the site. The Artesian Water Company, which
serves 131,000 people throughout New Castle County, has six wells
within three miles of the site. Further investigations are being
conducted.

NCR Corporation
Millsboro, Delaware
Waste water from an electroplating plant operated on the fifty-
eight-acre site from 1967 to 1974 by NCR was disposed of in an
unlined pit. Groundwater has, as a result, been contaminated with
hexavalent chromium, chloroform, TCE, and other toxics. The
drinking water of approximately 4,700 people who live within three
miles of the site has been polluted. TCE from the site enters the
Iron Branch (and then the Indian River) from a groundwater seep.
The site is being monitored and plans to clean it up are being
made.

FLORIDA

Harris Corporation
Palm Bay, Florida
Electronic devices and components are manufactured on the
500-acre site. A variety of volatile organic chemicals from the
operation have been found in water drawn from a well field supplying more than 18,000 Palm Bay residents, and heavy metals have
been found in groundwater underlying the site. Past spills are the
suspected origin of the contaminants. Harris Corporation started
pumping water, treating it (by aeration), and returning it to the
aquifer in 1985.

Peak Oil Company/Bay Drum Company
Tampa, Florida
Between the late 1950s and the late 1970s, Peak Oil recycled

waste oil and Bay Drum recycled metal drums on the site. Wastes from both operations were disposed of in lagoons on the site. Water both on the surface and in the ground and soil on the site is contaminated with pesticides, heavy metals, solvents, and PCBs. The Brandon Well Field, which is part of the Hillsborough County Water Supply System, is less than two miles from the site. The system supplies drinking water to about 57,000 people.

Petroleum Products Corporation
Pembroke Park, Florida
The Biscayne Aquifer supplies water to 150,000 people. Area wells are being contaminated with lead and PCBs, the result of an oil-recycling operation run on the site from 1952 to 1972. Spills and equipment maintenance are suspected to have been the primary sources of contamination. The two-acre lot was filled and paved over in 1974 in an unsuccessful attempt to curb groundwater pollution. In 1984 the Florida Department of Environmental Regulation filed suit to force the company to clean up the site. The cleanup was completed by 1987. The city of Hallendale has a well field within half a mile of the site, and two other city wells are located within three miles.

Sydney Mine Sludge Ponds
Brandon, Florida
The ponds were used to store wastes from a now-abandoned 1,700-acre phosphate strip mine, and industrial wastes and sludge from septic tanks were also dumped into them. Toluene, benzene, and solvents from the pond have contaminated the Hawthorne Formation, an aquifer that provides drinking water to 4,000 people within three miles of the site. Turkey Creek, which passes a half mile east of the ponds, has been contaminated with cadmium, zinc, chromium, and lead from the ponds. The county has constructed an underground slurry wall around the ponds to slow the escape of contaminants, and is attempting to determine the best way of cleaning up the site.

GEORGIA

Olin Corporation
Augusta, Georgia
Chlorine and caustic soda are manufactured at the plant. About 32,000 tons of mercury-contaminated wastes from the process have been disposed of in three waste pits on the site, and groundwater in the area has become contaminated with mercury as a result. Eleven Richmond County drinking water wells are located within three miles of the site. The extent of contamination is being studied, and the site has been proposed for addition to the NPL.

Robins Air Force Base
Houston County, Georgia
The 8,800-acre base is located east of Warner Robins, Georgia. There are a total of thirteen hazardous waste dumps on the base, one of which (the forty-five-acre sludge lagoon at Landfill #4) has been put on the priority cleanup list. An aquifer that supplies drinking water to more than 10,000 people is being contaminated with the solvents TCE and tetrachloroethylene, and surface water on the base has been found to contain phenols. The site is being studied for future cleanup.

HAWAII

Kunia, Waipahu, Waipio Heights, and Mililani Well Fields
Island and County of Oahu, Hawaii
Drinking water wells on the Schofield Plateau and the Ewa Plain operated by the City and County of Honolulu on Oahu have become contaminated with pesticides, including trichloropropane (TCP) and the nematocides ethylene dibromide (EDB) and di-bromochloropropane (DBCP). A total of six contamination sites have been proposed for addition to the NPL. It is the first time contamination resulting from the use of registered pesticides has resulted in such an action on the part of the EPA. The city has been filtering the water from the wells with granular activated

carbon and passing it through aeration towers to remove the contaminants.

ILLINOIS

Joliet Army Ammunition Plant
Joliet, Illinois
More than four billion pounds of explosives were produced between the early 1940s and 1977 on the fourteen-square-mile site, located south of Joliet (west of Illinois State Highway 53). Jackson Creek and Grant Creek, which flow through the site, and underlying groundwater were severely contaminated by waste water and chemical spills from the installation. A large quantity and variety of toxics are involved. Investigation into the nature and extent of the contamination is under way.

Outboard Marine Corporation
Waukegan, Illinois
A total of nine million pounds of PCBs were used in aluminum die-casting machinery on the site between 1959 and 1971. In 1976, it was found that PCBs leaking from the machines had been discharged into a ditch behind the plant, and had then made their way into Waukegan Harbor on Lake Michigan. Sediments on the bottom of the ditch and the harbor are contaminated with PCBs. The plant is believed to be one reason for the high concentrations of PCBs found in fish in the lake. The site is Illinois' highest priority NPL site.

Parsons Casket Hardware Company
Belvidere, Illinois
An electroplating process was used to finish metal fittings used in the production of caskets on the two-acre site from 1898 until the company went bankrupt in 1982. Waste water containing high concentrations of lead, cyanide, copper, and nickel was dumped in an unlined lagoon, which has polluted underlying groundwater. Leaking drums and tanks holding toxic wastes have also contributed to the contamination. Drinking water for Belvidere's residents

comes from wells that are between a quarter-mile and three miles from the site, which is being studied for future cleanup.

Stauffer Chemical Company
Chicago Heights, Illinois
Arsenic, antimony, and selenium have contaminated a shallow aquifer under a waste dump used from 1958 to 1979 to dispose of hazardous wastes associated with the production of pesticides and other chemicals. Wells that provide drinking water to 70,000 people are located between one and three miles from the site. Further study of groundwater contamination at the site is being conducted.

U.S. Ecology, Incorporated
Sheffield, Illinois
From the late 1960s until it was closed in 1983, the forty-five-acre site was the largest hazardous waste disposal facility in Illinois. Wastes containing pesticides, heavy metals, acids, chlorinated hydrocarbons, solvents, and PCBs were accepted. Many of the toxics dumped there have been discovered in the shallow aquifer underlying the site, and it has been proposed for addition to the NPL. Monitoring to determine the extent of contamination is being conducted by U.S. Ecology.

INDIANA

Midwest Solvent Recovery Company, Incorporated
Gary, Indiana
A solvent-recovery and industrial waste disposal facility was operated on the seven-acre site, located across the highway from the Gary Airport, until a fire in 1977 consumed most of the facilities located above the ground. The company abandoned the site without cleanup. Groundwater, soil, and possibly surface water in the area were contaminated with solvents and hazardous chemicals. In 1985, the EPA used CERCLA emergency cleanup funds to remove more than 85,000 drums and 5,000 cubic yards of severely contaminated soil. Almost half a million people live within three miles of the site.

Southside Sanitary Landfill
Indianapolis, Indiana
Heavy metals from the landfill, which has operated on the site from 1971 to the present, have contaminated one of the most productive aquifers in the state, and nearby Eagle Creek. Ways of stopping the contamination are currently being studied.

Waste Incorporated Landfill
Michigan City, Indiana
The landfill operated from 1966 to 1982 on ten acres of wetlands inside Michigan City (34,000 people live in the area). Hundreds of thousands of tons of industrial wastes heavily laced with PCBs, heavy metals, and solvents were buried there. The site, which started out at ground level, now stands fifty feet above the surrounding terrain. Heptachlor and other chemicals have been detected in groundwater under the site, and contamination also enters Trail Creek, which empties into Lake Michigan. A study of the extent of contamination is under way.

IOWA

Aidex Corporation
Council Bluffs, Iowa
A plant where pesticides were mixed operated on the site until it burned down in 1976. The 100,000 gallons of water used to fight the fire contaminated soil and groundwater on the site. A large underground waste storage tank may also have contributed to the contamination. During 1983 the U.S. Army Corps of Engineers coordinated a cleanup of the most severely polluted soils and liquid wastes still on the site. Investigation into the extent of contamination and ways of cleaning it up is under way.

Chemplex Company
Clinton, Iowa
Chemplex has manufactured polyethylene at the location since 1968. Peroxides, mineral spirits, vinyl acetate, styrene, benzene,

toluene, and polyaromatic hydrocarbons were contained in wastes dumped in an unlined landfill and waste lagoon on the site. Many of these toxics have also been discovered in nearby aquifers. The company has installed a system to recover and clean part of the polluted groundwater. The EPA has proposed that the site be added to the NPL.

Lawrence Todtz Farm
Camanche, Iowa
The site is composed of six acres of abandoned gravel pits into which municipal and industrial wastes were dumped between 1958 and 1975. An estimated 4,300 tons of liquid wastes from a DuPont cellophane plant in Clinton are among the industrial wastes dumped in the pits. Plasticizers (phthalates) and other hazardous substances have contaminated the Mississippi Alluvial Aquifer, which underlies the site. About 6,000 residents use water drawn from the aquifer within three miles of the site.

KANSAS

National Industrial Environmental Services
Furley, Kansas
The 160-acre site, located ten miles northeast of Wichita, started operation as a hazardous waste landfill in 1977. The dump was closed in 1982 when it was discovered that groundwater, surface water, and soil around the installation were contaminated with a variety of toxics. The old, leaking landfill has been closed and capped, and surface drainage ditches and a new landfill have been constructed on the site to slow the spread of contamination. The site has been proposed for addition to the NPL.

Obee Road
Hutchinson, Kansas
The city's landfill is the apparent source of a plume of groundwater pollution in the vicinity of Obee Road in east Hutchinson. Benzene, chlorobenzene, vinyl chloride, and toluene have been found in the shallow aquifer underlying the dump. A new well was

drilled for the Obee School District in 1987 when its old one was found to be contaminated. The extent of pollution at the site is being studied.

KENTUCKY

A. L. Taylor
Brooks, Kentucky
The roughly 17,000 drums of toxic wastes discovered there in the late 1970s gave the site, located twelve miles south of Louisville, the nickname "Valley of the Drums." A chemical dump, which handled primarily wastes from the paint and coatings industry in Louisville, was operated there from 1967 to 1977. The EPA and some of the companies that had dumped hazardous materials at the dump removed the drums between 1979 and 1981. Groundwater under the site is tainted with a variety of toxic chemicals. The extent of contamination and means of cleaning it up are being studied.

Howe Valley Landfill
Howe Valley, Kentucky
The landfill, located four miles south of town, was used between 1967 and 1976. Arsenic, chromium, and other metals are among the toxics disposed of in the dump, which is located in a ten-acre sinkhole. Groundwater threatened with pollution from the site includes that used by twenty-five residences within a mile of the site, and a spring two miles away that supplies water for 35,000 people. No cleanup effort has yet been initiated.

LOUISIANA

Dutchtown Treatment Plant
Ascension Parish, Louisiana
Chloroform, benzene, ethylbenzene, carbon tetrachloride, toluene, and industrial solvents have been found in the shallow

aquifer underlying the site, which is a quarter of a mile from
wetlands associated with the Mississippi River. The owner reclaimed oil on the site from 1965 to 1984, and dumped wastes into a
holding pond. Further investigation of the extent of contamination
is planned.

MAINE

McKin Company
Gray, Maine
The ten-acre former gravel pit was first used for waste disposal
when the Norwegian tanker *Tomano* ran aground in Hussey Sound,
spilling 100,000 gallons of industrial fuel. Septic-tank sludge and
industrial waste were later dumped in the pit. In 1978, the state
removed hazardous materials from the surface of the site after
receiving numerous complaints from residents of the area who said
their water smelled bad and stained laundry. Further investigation
into the extent of contamination and ways of cleaning it up is in
progress.

MARYLAND

Aberdeen Proving Ground
Edgewood, Maryland
Since 1917 development and testing of chemical weapons has
been conducted at the 79,000-acre installation (located near the
head of Chesapeake Bay). Significant quantities of napalm, white
phosphorus, arsenic, cyanide, and chemical agents used in nerve
gas have been disposed of on the grounds over the years. The state
recommended in 1983 that three wells in the Canal Creek area near
the base be closed because of contamination. The Long Bar Harbor
well field of the Harford County Department of Public Works and a
well field used by the Joppatowne Sanitary Subdistrict, which
together serve an estimated 35,000 people, are within three miles of
the facility. Investigation into the location of wastes buried on the

proving grounds and of the extent and pattern of groundwater pollution on the site is continuing.

Mid-Atlantic Wood Preservers
Harmans, Maryland
Lumber is pressure-treated with a preservative (chromated copper arsenate) on the three-acre site. Arsenic and chromium have contaminated groundwater under the site. Storage tank overflows and drips from the treated lumber appear to be the source of the contamination. An estimated 75,000 people's drinking water comes from wells within three miles of the site. The extent of contamination and alternatives for cleaning it up are being reviewed.

Woodlawn County Landfill
Woodlawn, Maryland
The county operated a landfill on the site, an abandoned gravel quarry, twenty-four hours a day, from 1965 to 1979. Agricultural, municipal, and industrial wastes were dumped there. An estimated 783 *tons* of polyvinyl chloride (PVC) were disposed of at the landfill by the Firestone Tire and Rubber Company. PVC, benzene, toluene, and lead have been found in groundwater under the site. Private wells between 400 feet and three miles of the dump supply drinking water to 5,700 people. No cleanup is as yet under way.

MASSACHUSETTS

Baird and McGuire
Holbrook, Massachusetts
Pesticides, disinfectants, floor strippers and waxes, coal-tar emulsions, and related products have been formulated on the site since 1912. Arsenic, creosote, and a variety of toxic organic chemicals have contaminated underlying groundwater. Holbrook's South Side Well Field, which has been abandoned because of contamination, is within 1,000 feet. The Cochato River, which is 500 feet from the property, is diverted two-and-a-half miles downstream into the Richardi Reservoir, the source of drinking water for 90,000 people in Braintree, Holbrook, and Randolph. After surface runoff

caused an oil slick on the river in 1983, the EPA took steps to stabilize the toxic wastes on the site. The extent of contamination and possible ways of cleaning it are being studied.

Industri-plex
Woburn, Massachusetts
Insecticides, explosives, acids, and other industrial chemicals were manufactured by Merrimac Chemical Company, and later by Monsanto Company, on the site between 1953 and 1981. In addition to the wastes left behind by the two chemical companies, a variety of other toxic industrial compounds accumulated in soils and groundwater under the property in more than 130 years of use. The extent and nature of contamination that has resulted is still being studied.

MICHIGAN

Barrels, Incorporated
Lansing, Michigan
The 1.8-acre site, which is located inside Lansing's city limits, was used in a barrel-recycling operation from 1964 to 1981. The residues left in the barrels, including a wide range of toxic chemicals, were dumped on the ground, and extensive contamination of the shallow aquifer under the site has resulted. The deeper Saginaw Formation, the aquifer that supplies drinking water to Lansing's 133,000 residents, is connected to the contaminated aquifer. The Michigan Department of Natural Resources removed contaminated barrels and soil and pumped the toxic contents from underground storage tanks in 1986 in an attempt to slow the flow of contaminants from the site. Further investigation into the extent of groundwater pollution near the site is under way.

Berlin and Farro
Swartz Creek, Michigan
The company operated an incinerator for liquid industrial wastes on the site from 1971 to 1980. A landfill holding an estimated 10,000 crushed drums, underground tanks containing approx-

imately 30,000 gallons of C-56 (an extremely toxic by-product of the pesticide manufacturing process), and four lagoons holding about 11,000 gallons of severely contaminated waste sludge were believed to be the primary sources of extensive groundwater pollution near the site. By 1983, most of these toxics had been removed, and a temporary cap had been placed over the waste lagoons. Further study of the extent of contamination caused by the site and ways of cleaning it up is in progress.

Rockwell International Corporation
Allegan, Michigan
Universal joints for automobiles have been manufactured at the site since 1910. Between 1910 and 1960, wastes from the operation were discharged into the nearby Kalamazoo River. Between 1960 and 1972 the wastes were discharged, untreated, into two ponds on the thirty-acre site. In 1972 a treatment plant was built to process the wastes before discharge into the waste ponds. Lead, arsenic, cyanide, and a variety of other toxic materials have shown up in nearby groundwater and in the Kalamazoo River. The cleanup effort is in the planning phase.

MINNESOTA

Reilly Tar
St. Louis Park, Minnesota
The company operated a coal-tar distillation and wood preservative plant on the site from 1917 to 1972. Wastes from the processes were dumped on-site and into a series of nearby ditches (see description in chapter 4). Benzene, pyrene, and a variety of poly-aromatic hydrocarbons associated with coal tar have contaminated an aquifer used by a quarter of a million people to a depth of 900 feet. The pollution plume in the aquifer reaches two miles from the site. The EPA sponsored the removal of some of the contaminants still on the site's surface and the plugging of wells that were facilitating the migration of toxics in the early 1980s. Further study of the extent of the contamination associated with the site and ways of controlling further pollution is under way.

St. Augusta Sanitary Landfill/Engen Dump
St. Augusta Township, Minnesota
Benzene, arsenic, lead, chromium, and TCE have contaminated the shallow aquifer under the fifty-acre site, which is just west of the Mississippi River, as the result of waste dumping between 1966 and 1983. Paint wastes, solvents, and ashes from the incineration of hazardous wastes were dumped there. Further investigation into the extent of contamination is being conducted.

Twin Cities Air Force Reserve Base
Minneapolis, Minnesota
A landfill used between 1963 and 1972 at the base, which is located adjacent to and within the cities' airport, has contaminated the underlying aquifer. An estimated 67,400 people rely on wells that are within three miles of the dump for drinking water. Contaminants from the dump are washed into the nearby Minnesota River during periods of high water. Paints, primers, lacquers, paint thinners and removers, and sludge from leaded fuel were dumped at the site.

University of Minnesota Rosemount Research Center
Rosemount, Minnesota
UM used the four-acre dump to dispose of chemical laboratory wastes from 1960 to 1973. Heavy metals and a variety of hazardous chemicals have contaminated the aquifer underlying the site. Several private wells have been polluted, and the University is supplying twenty-eight Rosemount families with bottled water as a result. About 9,600 people use water from wells located within three miles of the site.

MISSOURI

Conservation Chemical Company
Kansas City, Missouri
A facility for the treatment and disposal of hazardous wastes was operated on the site, located at the confluence of the Blue and Missouri Rivers, from 1960 to 1980. An estimated 300,000 tons of

wastes, primarily from the metal-finishing industry, were accepted for disposal. Most were dumped in unlined lagoons. Contaminants from the dump have entered local groundwater and the Missouri River. Further study of how best to clean up the installation is in progress.

Weldon Springs Quarry
St. Charles County, Missouri
The Army operated an explosives plant on the site, which is about thirty miles west of St. Louis, from 1941 to 1944. Hazardous materials left over from the plant were disposed of in an old limestone quarry on the grounds. The Atomic Energy Commission dumped uranium, thorium, and their decay products, which came from an uranium-enrichment operation on the site, into the pit from 1959 to 1969. Groundwater and surface water in the area have become contaminated with radioactive materials as a result. An estimated 58,000 people in St. Charles County use water from a well field located half a mile from the quarry.

MONTANA

Silver Bow Creek/Butte Area
Butte, Montana
A century of extensive mining and refining of copper ores in Butte and nearby Anaconda have caused extensive contamination of Silver Bow Creek, one of the Clark Fork River's headwaters. Most of the ground and surface water in the area has become contaminated, and the burden of contamination has also been carried hundreds of miles down the Clark Fork. Some of the mine and smelter tailings that are one source of the contamination are being removed, and the extent of contamination and ways of minimizing future water pollution are being studied.

Montana Pole and Treating
Butte, Montana
A plant that treated fence posts, utility poles, and bridge timbers with a mixture of 95 percent diesel oil and 5 percent pen-

tachlorophenol (PCP) was run by the company on the forty-acre site from 1947 to 1983. Thousands of gallons of the mixture, which also contained traces of dioxin, soaked into the ground as the result of spills and leaking storage tanks over the years, thoroughly fouling the underlying aquifer. The contamination was discovered when polluted water from the site was found to be polluting nearby Silver Bow Creek (see review above). A Coast Guard emergency response team sacked up approximately 10,000 cubic yards of soil that was severely contaminated with the PCP/dioxin mix in 1986 and stored it in buildings on the site, where it will remain until a viable means of detoxifying it is available. Pumps removed more than 20,000 gallons of diesel oil laced with PCP and dioxin from the aquifer during their first year of operation.

NEBRASKA

Cornhusker Army Ammunition Plant
Hall County, Nebraska
Bombs, artillery shells, rocket boosters, and land mines were produced at the nine-square-mile installation between 1942 and 1973. Cesspools, landfills, and incinerators located throughout the site were used to dispose of toxic wastes including trinitrotoluene (TNT) and the experimental explosive, RDX. The Army paid for an extension of the municipal water system from Grand Island, which is three miles west of the site, to replace individual wells that became contaminated as a result. More than 500 private wells so far have been polluted by toxics seeping out of the site. The Army has started to incinerate explosive-contaminated soil in the area and is studying ways of cleaning up the polluted groundwater.

NEW HAMPSHIRE

Keefe Environmental Services
Epping, New Hampshire
Laboratory wastes, waste oils, paint sludges, toxic organic chemicals, and heavy metals were accepted at the fifty-acre waste-

storage and transfer station from 1978 to 1981. Spills, leaking drums, and a 750,000-gallon storage lagoon caused extensive contamination of soils with toxics. In 1980 and 1981 the EPA used $1 million in CERCLA emergency cleanup funds to remove part of the wastes and to slow the leakage of the remaining materials. The site is being further studied to assess the extent of contamination and ways of cleaning it up.

NEW JERSEY

Curcio Scrap Metal, Incorporated
Saddle Brook Township, New Jersey
Electrical transformers were among the scrap metal recycled at the company's yard, and PCBs that poured into the ground when the transformers were cut up have contaminated surface runoff, underlying soil, and groundwater. The Brunswick Formation, New Jersey's most important and extensive aquifer, is threatened with contamination as a result. An estimated 93,000 people use water drawn from wells within three miles of the site. No cleanup measures have yet been taken.

Dayco Corporation/L. E. Carpenter Company
Wharton Borough, New Jersey
Approximately 20,000 gallons of solvents were estimated by a consultant to be floating on the surface of the groundwater under the site in 1983. Drums of polyvinyl chloride (PVC) were buried there until 1970. In 1982 the company removed 4,000 cubic yards of tainted soil and PVC sludge in an attempt to stop the spread of pollution. The Quarternary Aquifer, the only source of drinking water for the Rockaway River Basin, is being polluted by the toxics. Municipal wells serving 27,000 people in Wharton Borough and Dover Township are located within three miles of the site. Further analysis of the extent of the pollution coming from the grounds is under way.

Lipari Landfill
Pitman, New Jersey
A landfill that accepted municipal and industrial wastes was

operated in the six-acre former gravel pit from 1958 to 1971. A variety of toxic industrial chemicals polluted the aquifer under the site and seeped into the nearby Chestnut Branch of Rabbit Run and Alcyon Lake. The EPA spent nearly $3 million in the early 1980s assessing the contamination caused and partially cleaning it up. An underground slurry wall and a bentonite cap are to be constructed to further reduce the contamination caused by the site. The Lipari Landfill is number one on the NPL.

NEW YORK

Conklin Dumps
Conklin, New York
The two landfills, which cover a total of 619 acres, were used by the Town of Conklin from 1964 to 1975. Groundwater under the site has been found to contain high concentrations of arsenic, chromium, mercury, benzene, and a variety of other toxics, as do a number of nearby private wells. An estimated 2,000 people use water from wells within three miles of the site. A large wetland is adjacent to the area. Studies of how best to deal with the contamination are being conducted.

Genzale Plating Company
Franklin Square, New York
Waste water containing chromium, copper, nickel, and zinc were disposed of into three leaching ponds on the half-acre site starting in 1915. In 1983, in response to a 1981 order by the Nassau County Health Department, the company removed the sludge in the ponds and some of the heavy-metals-contaminated soil to an approved toxic waste disposal site. A well for the Franklin Water District, which serves 20,000 people, is located 1,700 feet from the site, and wells for the West Hempstead-Hempstead Gardens Water District, which deliver drinking water to another 32,000 people, are within three miles of the site. The extent of contamination and possible means of cleaning it up are being studied.

Goldisc Recordings, Incorporated
Holbrook, New York
Goldisc (formerly Sonic Recording Products) produced pho-
nograph records on the site from 1968 to 1983. Wastes disposed of
there include solvents, hydraulic oil, and nickel-plating materials.
The underlying aquifer, which is threatened with contamination as
a result of the waste dumping, is the only source of drinking water
in the area. An estimated 130 wells located within three miles of
the site provide water to more than 71,000 people. A public water
supply well is located within a 1,000-ft. down-gradient. Further
study of the extent of contamination and of ways of cleaning it up is
in progress.

Jones Sanitation
Hyde Park, New York
Sewage sludge and industrial waste were disposed of in unlined
pits on the Dutchess County site from 1956 to 1979. Solvents,
metals, phenols, methylene chloride, chloroform, and naphthalene
were among the toxic materials that seeped into the pits, some of
which were deeper than the surface of the groundwater, which is
five to seven feet below ground. At least twenty-three wells serving
9,500 people are located between 1,000 feet and three miles from
the site. Surface water in a nearby wetland is also threatened.
Options for restoring the site are being considered.

Love Canal
Niagara Falls, New York
Hooker Chemical's 16-acre hazardous-waste dump is undoubtedly the best known in the country (described in chapter 1).
Hooker and others disposed of an estimated 20,000 tons of toxic
chemicals at the dump between 1947 and 1953. Between 1977 and
1980 the State of New York and the federal government together
spent $45 million in an attempt to minimize the impact on human
health caused by the dump ($30 million went for health testing and
the relocation of area residents). In 1982, the EPA authorized $7
million for the state to study the site further, and eventually build
an underground slurry wall around it and a cap over it.

Richardson Hill Road Landfill/Pond
Sidney Center, New York
A twenty-foot by eighty-foot pit on the six-acre site (on the west side of Richardson Road) was used between 1963 and 1970 by the Bendix Corporation's Electronic Components Division for the disposal of waste oils, equipment, and parts. Solvents and vinyl chloride from the landfill have been found in area groundwater and surface water. Allied Corporation, which now owns Bendix, is buying bottled water for individuals whose wells have been contaminated as a result of the landfill. Plans are being made for eventual cleanup of the site.

Rowe Industries
Noyack/Sag Harbor, New York
A plume of groundwater pollution 500 feet wide, 2,600 feet long and 12 to 124 feet deep had seeped out from the site by 1986. Wells serving fifteen homes had already been contaminated, and another thirteen were threatened at the time. Approximately 3,500 residents with private wells and 2,500 customers of the Suffolk County Water Authority use groundwater coming from wells located within three miles of the site. In 1985, the EPA used CERCLA emergency funds to extend public water supply mains to twenty-five affected homes. A study of possible ways of cleaning up the water is under way.

NORTH CAROLINA

Carolina Transformer Company
Fayetteville, North Carolina
The company has recycled electrical transformers at the site since 1958, and water both on the surface and in the underlying aquifer has become contaminated with PCBs and chlorobenzenes as a result of spills. In 1984, the EPA ordered Carolina Transformer to clean up the site, but the company refused. The EPA then took responsibility for the cleanup and removed almost 1,000 tons of contaminated soil to a certified hazardous waste landfill. The

Department of Justice has filed an action against the company, seeking to recover the cost of the cleanup, plus damages for failure to respond to EPA's order that the site be cleaned up.

Charles Macon Lagoon and Drum Storage
Cordova, North Carolina
The sixteen-acre site, located a mile and a half south of Cordova on State Road 1103, was used as a waste oil recycling facility until 1981. Industrial solvents, acids, and bases from other companies were also disposed of there. By 1983, all the more than 2,000 drums of chemical wastes and ten on-site waste lagoons had been cleaned up, primarily through the use of CERCLA emergency funds. In 1985, barium, chromium, and toluene had been found to have migrated downstream from the site through the underlying aquifer.

OHIO

Industrial Excess Landfill
Uniontown, Ohio
Residential and commercial wastes and an estimated one million gallons of industrial wastes were dumped at the landfill while it was in operation from 1959 to 1980. Low levels of several hazardous chemicals that leaked from the facility have been detected in nearby private wells. The extent of pollution caused by the landfill is currently being studied.

TRW, Incorporated
Minerva, Ohio
Metal-casting operations have been conducted on the fifty-four-acre site since 1954. PCBs, TCE, TCA, and a variety of other toxics used at the plant were stored in drums (which eventually leaked) and in a waste pond, and have been found in groundwater under the site. Minerva's city wells (serving about 4,500) are located one mile southwest of the installation in the direction of groundwater flow. During 1985, TRW, at the state's request, cleaned up soil, sediments, and wastes remaining on the site. An effort to remove contaminants from area groundwater is now under way.

OREGON

Allied Plating Company
Portland, Oregon
The chrome-plating company operated on the site, 600 feet south of the Columbia River in northern Multnomah County, from 1957 to 1984. Wastes were discharged, untreated, into a half-acre pond behind the plant that had formerly been a swamp. Chromium and barium have contaminated groundwater under and near the site as a result. Private and public wells located within three miles supply water to 1,500 people. The site is being studied to assess the degree of contamination and potential ways of cleaning it up.

Martin Marietta Aluminum Company
The Dalles, Oregon
Both shallow and deep aquifers underlying the company's 350-acre site in The Dalles are contaminated with cyanide, a by-product of the aluminum smelting process. The company removed 75,000 tons of cyanide-bearing wastes (when ordered by the state to do so) to prevent further contamination. Approximately 14,000 people in The Dalles and nearby Chenoweth use groundwater for domestic purposes.

PENNSYLVANIA

American Electronics Laboratories
Montgomeryville, Pennsylvania
Electronic communications equipment is made at the plant. Solvents including TCE and TCA, which are disposed of on the grounds, have been found in nearby public and private wells. One public drinking water well is within fifty feet of the installation. An estimated 106,000 people use water from wells located within three miles of the site. The company has removed some of the most seriously contaminated soils to a hazardous waste dump, and is pumping polluted water from on-site monitoring wells for treatment and disposal at a nearby sewage plant.

Butler Mine Tunnel
Pittston, Pennsylvania
The tunnel was constructed about fifty years ago to drain an estimated five square miles of underground coal mines. It empties directly into the Susquehanna River. During July, 1979, an oil slick formed on the river, starting at the tunnel and extending sixty miles downstream to the water intakes for Danbury, Pennsylvania. The owners of a local service station and a waste disposal company were jailed after it was proven they had caused the contamination by dumping hazardous wastes down a well that led into the old mines. Between 1979 and 1980, barriers installed on the river by the EPA to stop the spread of the pollution collected about 160,000 gallons of oil laden with hazardous chemicals. No further pollution issued from the mine until September, 1985, when heavy rains associated with Hurricane Gloria flushed another 100,000 gallons of contaminated waste oil into the river. Further study of the extent of contamination at the site and ways of cleaning it up is under way.

C & D Recycling
Foster Township, Pennsylvania
Lead-cased telephone cables were recycled on the site between 1960 and the early 1980s. The cables were burned to melt off the lead so the copper could be reclaimed. Lead and copper from the burning areas and disposal pits have contaminated the underlying aquifer. Approximately 6,100 people rely on wells located between 1,000 feet and three miles from the site. The extent of contamination at the site is now being surveyed, and plans are being made for an eventual cleanup.

Letterkenny Army Depot
Chambersburg, Pennsylvania
The maintenance, overhaul, and rebuilding of vehicles and missiles since 1947 has produced large quantities of spent solvents that have been disposed of in a dump at the base's southeast corner. About forty residential wells as much as two-and-a-half miles east of the base have become contaminated with solvents as a result. The Army paid to have the homes hooked up to a public water system in 1987. An investigation of the extent of the contamination and of possible cleanup methods is being conducted.

Whitmoyer Laboratories
Jackson Township, Pennsylvania
Pharmaceuticals for animals were produced on the 17.5-acre site from 1934 to 1984. Arsenic and a variety of other hazardous waste chemicals generated by the operation were disposed of in unlined lagoons. The company provided bottled water to neighboring residents starting in the mid-1960s, when it was discovered that their wells were contaminated with arsenic and other toxics. A study of the extent of contamination and possible ways of cleaning it up is in progress.

RHODE ISLAND

Central Landfill
Johnston, Rhode Island
The 133-acre dump now handles only municipal waste, but hazardous industrial wastes were accepted in the past (the state estimates 1.5 million gallons of hazardous wastes were disposed of at the dump during 1978 and 1979). The areas where hazardous materials were disposed of have been closed and cleaned up (by state order), but soil and groundwater under the site are still contaminated, primarily with solvents.

SOUTH CAROLINA

Sangamo-Weston, Incorporated
Pickens, South Carolina
Electrical capacitors have been manufactured on the site since 1955. PCBs were used in the capacitors until 1976. Nearly 40,000 cubic yards of PCB-tainted wastes were dumped on the site. Surface streams, nearby Lake Hartwell, and groundwater in the area have become contaminated with PCBs as a result. The Easley-Central Water Plant, which supplies 14,500 people with water from Twelve-Mile Creek, has also been contaminated. The company has removed part of the wastes and contaminated soil from the grounds.

TENNESSEE

Arlington Blending and Packaging Company
Arlington, Tennessee
Pesticides were blended and packaged on the site between the 1950s and 1979. Aldrin, dieldrin, endrin, chlordane, heptachlor, lindane, and methyl parathion were among the toxic chemicals handled. About 1,200 fifty-five-gallon drums and deteriorating bags of pesticides were left behind when the company went out of business. The EPA used CERCLA emergency cleanup funds to remove 3,500 gallons of chemicals and 1,920 cubic yards of contaminated soils. Underlying the site are three layers of water-bearing strata that are used for drinking water in the area. Two community water-well systems between 1,200 feet and three miles away (serving Arlington and nearby Galloway) provide drinking water to an estimated 2,700 people.

Milan Army Ammunition Plant
Milan, Tennessee
Outmoded munitions were dismantled at the plant between 1942 and 1978. Explosives such as TNT and RDX were removed from the weapons by exposing them to a high-pressure stream of hot water. After the water had absorbed the chemicals it was dumped into one of eleven unlined lagoons on the base. The explosives and heavy metals have contaminated groundwater under the site. The same aquifer is used to supply more than 13,000 people with drinking water within three miles of the site. Plans for further monitoring of the extent of contamination and eventual cleanup are being made.

TEXAS

Bailey Waste Disposal
Bridge City, Texas
The ten-acre site north of the Neches River, two miles southwest of Bridge City, is contaminated with chloroform, phthalates, TCE,

and other hazardous chemicals—the result of approximately 72,000 cubic yards of industrial wastes being dumped there between 1950 and mid-1960. About 7,600 people rely on wells within three miles of the site as their primary source of drinking water. The state is conducting a study of the extent of contamination and possible ways of cleaning it up.

Sol Lynn/Industrial Transformers
Houston, Texas
The one-acre site is half a mile from the Houston Astrodome and Astroworld amusement park. TCE and PCBs used in the production of transformers on the site have severely contaminated underlying groundwater. Further investigation into the extent of the pollution caused by the dumping is being conducted.

Stewco, Insorporated
Waskom, Texas
Stewco is a company that formerly hauled glue, resin, creosote, and fuels for oil companies. The tank trucks used for the hauling were steam-cleaned between runs with an alkaline solution, which drained into a series of unlined evaporation ponds after use. Soil and water in and under the site are contaminated with methylene chloride, naphthalene, phthalates, toluene, DDT, arsenic, mercury, lead, cadmium, and other toxics. The EPA used emergency cleanup funds to remove the liquids and sludges remaining in the ponds, and transport them to a hazardous waste dump. Further study of the extent of contamination at the dump is under way.

UTAH

Midvale Slag
Midvale, Utah
A smelter that processed copper, gold, silver, and lead ores on the 300-acre site (which is within the Salt Lake City metropolitan area) was run by the United States Smelting, Refining, and Mining Company from 1902 to 1971. Heavy metals in the remaining two

million tons of slag threaten underlying groundwater. Several municipal wells that serve about 38,000 people are located within three miles of the site. A preliminary assessment of the site is under way.

Sharon Steel Corporation
Midvale, Utah
Metals were milled on the 260-acre site from 1910 to 1971, creating an estimated ten million tons of tailings containing high concentrations of lead, arsenic, copper, cadmium, chromium, and zinc. Groundwater under the site is contaminated with arsenic and lead. An estimated 500,000 people use water drawn from wells within three miles. An investigation into the extent of the contamination and alternatives for cleaning it up is in progress.

Silver Creek Tailings
Park City, Utah
A number of mining companies operated on the site between 1900 and 1930, leaving behind an estimated 700,000 tons of mine tailings. The tailings were reprocessed during the 1940s; acids and solvents were poured on them to reclaim silver. During the late 1970s and the early 1980s, 130 single-family homes and fifty apartments were built on the tailings. Lead, cadmium, and silver have contaminated surface water in the area, and the likelihood that groundwater has also been polluted is high. Approximately 10,000 people live within three miles of the site and use groundwater.

Wasatch Chemical Company
Salt Lake City, Utah
Pesticides and other chemicals have been produced on the six-acre site since the early 1960s. An estimated 2,300 cubic yards of hazardous chemical wastes have been disposed of there by Wasatch and others since that time. Pesticides and a variety of related chemicals have been found in underlying groundwater. Approximately 85,000 people live within three miles, and private wells within three miles supply drinking water to 60,000 people. The EPA discovered dioxin in drums, standing water, and soil while undertaking an emergency cleanup of the site in 1986. Further

investigation into the nature and extent of contamination on the site is being conducted.

VIRGINIA

Culpeper Wood Preservatives, Incorporated
Culpeper, Virginia
The company has treated wood on the twenty-acre site, located on the outskirts of town, since 1976. In 1981, 100,000 gallons of waste water laced with arsenic and chromium spilled from an impoundment, tainting neighboring surface water. The aquifer underlying the site is contaminated with arsenic and chromium. An estimated 2,000 people use water from wells within three miles of the installation. No action is currently being taken to clean up the site.

Greenwood Chemical Company
Newtown, Virginia
Specialty chemicals were manufactured on the fifteen-acre site between the mid-1940s and 1985. Many toxic materials (including an estimated one to ten metric tons of cyanide per year) were used. Waste water from the operation was disposed of in one of eleven unlined lagoons. Benzene, chlorobenzene, and TCE have contaminated groundwater under the site and nearby. Fish and cattle have been killed at times when the ponds overflowed into nearby Stockton Creek. Private wells situated between 600 feet and three miles of the lagoons furnish water to an estimated 1,600 people. A preliminary investigation of the extent of contamination is being conducted.

WASHINGTON

Midway Landfill
Kent, Washington
The sixty-acre landfill, an abandoned gravel pit, was operated by the City of Seattle between 1966 and 1983. Some industrial wastes

were accepted. The aquifer under the dump is contaminated with lead, arsenic, toluene, xylene, and other toxics. More than 10,000 people rely on wells within three miles of the dump for their drinking water. Monitoring of groundwater flow near the site revealed that methane gas from the landfill had penetrated the soils underlying nearby commercial and residential areas. Eleven families had to leave their homes until the threat of explosions was reduced. The city is studying ways of controlling the spread of contamination from the dump.

McChord Air Force Base
Tacoma, Washington
More than 500,000 gallons of hazardous substances have been used and disposed of on the base since 1940. Chloroform, benzene, arsenic, chromium, and mercury have contaminated underlying groundwater and Clover Creek, which flows through the area. Lakewood Water District and American Lake Gardens (a private development) operate wells that draw from the aquifer underlying McChord. A total of well more than 10,000 people get drinking water from wells tapping the aquifer within three miles of the base. Investigation into the extent of pollution and into potential ways of cleaning it up are under way.

Old Inland Pit
Spokane, Washington
Wastes from a nearby steel foundry have been disposed of in the ten-acre former gravel pit since 1976. Arsenic, cadmium, chromium, lead, acetone, methylene chloride, toluene, and TCE are among the hazardous substances in the unlined pit. The Spokane Valley-Rathdrum Prarie Aquifer, which is the sole source of drinking water for more than 30,000 people who live within three miles of the site, underlies the site. While no contamination has so far been detected, the area's highly permeable soils make movement of the toxics into the aquifer likely. Further investigation is under way.

WEST VIRGINIA

Mobay Chemical Corporation
New Martinsville, West Virginia
Chemical-production wastes have been disposed of on the property since 1950, resulting in severely contaminated soils and groundwater. Benzene, chlorobenzene, vinyl chloride, and other toxics have been found in the aquifer underlying the site. Further investigation into the extent of contamination is under way.

WISCONSIN

Algoma Municipal Landfill
Algoma, Wisconsin
The 7.5-acre landfill was used between 1969 and 1983, primarily for the disposal of municipal garbage. However, painting materials including lacquer thinner, primer, pigments, and polyvinyl acetate were also dumped there. Water under the site has been contaminated with benzene, methyl ethyl ketone, xylenes, arsenic, and cadmium. An estimated 5,000 people use drinking water drawn from wells within three miles. A study of the extent of pollution and of possible ways of cleaning it up is under way.

Hagen Farm
Stoughton, Wisconsin
The five-acre site southeast of Stoughton was a gravel pit before it was (illegally) used between 1950 and 1960 for the disposal of toxic wastes. The Wisconsin Department of Natural Resources discovered the dumping in 1982. Approximately 5,000 drums of waste material were observed, and groundwater under the site was found to be fouled with xylene, acetone, ethylbenzene, vinyl chloride, and a variety of other industrial solvents. Most of the water for the city's community water system, which serves 7,500, comes from wells within three miles of the site. In 1983, the Wisconsin Department of Justice filed a suit against Uniroyal, Incorporated, and Waste Management of Wisconsin, Incorporated, calling for the companies to investigate and clean up the site.

Index